EVOLUTION OF THE AIR CAMPAIGN PLANNING PROCESS AND THE CONTINGENCY THEATER AUTOMATED PLANNING SYSTEM (CTAPS)

DANIEL R. GONZALES

Prepared for the
UNITED STATES AIR FORCE

RAND

This report summarizes an examination of the air campaign planning process, including observation of how the process was conducted in recent exercises and a review of how the process was performed during the Gulf War. A number of suggested changes to the process are recommended that, in conjunction with changes to the Contingency Theater Automated Planning System (CTAPS), could improve the process significantly and reduce the time needed for production of the Air Tasking Order from 48 to 24 hours.

CTAPS capabilities were examined as a part of this study. The CTAPS 5.0x and planned 6.0 architectures were reviewed and suggestions presented that could enhance the operational capabilities of the system.

This report should be of interest to project managers and monitors of CTAPS and related programs, to those interested in the air campaign planning process, and to those responsible for developing Department of Defense or Air Force information system architectures.

This work is sponsored by the Air Force Deputy Chief of Staff, Plans and Operations, and was performed within the Force Modernization and Employment program of Project AIR FORCE.

PROJECT AIR FORCE

Project AIR FORCE, a division of RAND, is the Air Force federally funded research and development center (FFRDC) for studies and analyses. It provides the Air Force with independent analyses of policy alternatives affecting the development, employment, combat readiness, and support of current and future aerospace forces. Research is being performed in three programs: Strategy and Doctrine; Force Modernization and Employment; and Resource Management and System Acquisition.

In 1996, Project AIR FORCE is celebrating 50 years of service to the United States Air Force. Project AIR FORCE began in March 1946 as Project RAND at Douglas Aircraft Company, under contract to the Army Air Forces. Two years later, the project became the foundation of a new, private nonprofit institution to improve public policy through research and analysis for the public welfare and security of the United States—what is known today as RAND.

CONTENTS

During a large air campaign, the Joint Force Air Component Commander (JFACC) may be presented with two conflicting goals. On the one hand, he would like to employ air power in the most effective and well orchestrated way possible by producing an Air Tasking Order (ATO)—that includes all U.S. and allied air sorties for the next 24 hour time period—in a well ordered deliberate planning process. On the other hand, the JFACC will want to have the ability to change target priorities and attack new high-priority targets as quickly as possible. But if too many changes are made to the ATO in too short of a time, the planning process can be severely impacted potentially leading to chaos and a dramatic reduction in the effectiveness of the overall air campaign. A balance must be struck between these two goals to maximize the application of air power in large air campaigns.

During Operation Desert Storm, it was hard to balance these goals because of the difficulties encountered in producing and disseminating the ATO. Although an automation aid, the Computer Assisted Force Management System (CAFMS), which was used for these purposes, had numerous drawbacks. It was based on obsolete computer hardware and only with tremendous difficulty was it modified to support the large sortie rate of coalition air forces. Because of the large number of sorties in each ATO and CAFMS limitations, it was very difficult to adjust the ATO properly when new high-priority targets or target changes were added in the later stages of the ATO production process. And even though planners had almost 40 hours to plan air attacks and prepare the ATO, they struggled throughout the war to produce fully coordinated and deconflicted ATOs within that time period.

Significant difficulties were also encountered in obtaining, deconflicting, and properly incorporating order of battle and bomb damage assessment information into the ATO planning process. There were several reasons for these problems, but one important factor was the lack of automation support for storing and processing this type of information and for other vital planning activities. Finally, many units received the ATO late or with tremendous difficulty. Again there were several reasons for the latter problem, but a primary culprit was CAFMS.

Before Desert Storm, numerous programs were under way in the Air Force and the Department of Defense (DoD) to address many of these deficiencies. In the last few years, many of these programs were combined into an umbrella program in an effort to better coordinate these development activities. The purpose of this umbrella program is to develop a state-of-the-art force-level command, control, communications, computers, and intelligence (C4I) system, called until recently the Contingency Theater Automated Planning System (CTAPS), that can support all phases of ATO production, dissemination, and execution monitoring.[1]

In this report, we examine CTAPS and the air campaign planning process and propose modifications to both that will help remedy these deficiencies and dramatically increase the responsiveness of U.S. air forces in large conflicts. We examine the functionality of CTAPS and the structure of the air campaign planning process in detail in the body of this report. Below we provide a top-level description of the system and summarize our findings for improving CTAPS and the air campaign planning process.

CTAPS 5.0x

CTAPS is a complex automated support system that runs on a large networked set of computer work stations. Because of its ancestry and the evolutionary acquisition approach used in its development, CTAPS is a complex combination of applications that have been modified to run together with minimal interference in the same client-server computing environment. The four key applications used in the ATO production process in version 5.0x of CTAPS and the key data flows in the process are shown in Figure S.1. There are many more applications in CTAPS 5.0x; however, for simplicity we have included only the major ones in the figure (see Figures 5.1 and 7.1 for more detailed illustrations).

The ATO planning cycle starts with the collection of intelligence data needed for air campaign planning. The Intelligence Correlation Module (ICM) is used to correlate

RAND *MR618-S.1*

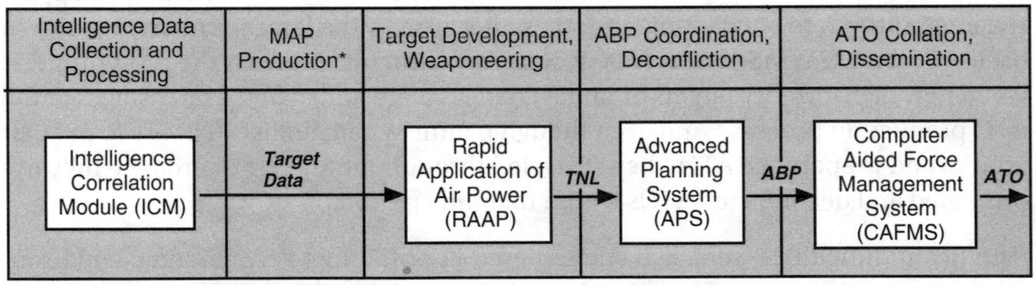

Intelligence Data Collection and Processing	MAP Production*	Target Development, Weaponeering	ABP Coordination, Deconfliction	ATO Collation, Dissemination
Intelligence Correlation Module (ICM)	*Target Data* →	Rapid Application of Air Power (RAAP) *TNL* →	Advanced Planning System (APS) *ABP* →	Computer Aided Force Management System (CAFMS) *ATO* →

TNL: Target Nomination List, MAP: Master Attack Plan, ABP: Air Battle Plan, ATO: Air Tasking Order.
*Manual Process if only CTAPS 5.0x is available.

Figure S.1—Key Applications and Data Flows in the CTAPS 5.0x ATO Production Process

---

[1]CTAPS has recently been renamed the Theater Battle Management Core System (TBMCS). Since much of the analysis in this report concerns specific software versions of CTAPS, for simplicity we refer to future versions of TBMCS as CTAPS.

order of battle, situation awareness, and target intelligence information received from external sources. In many CTAPS configurations, ICM supplies target data—including the targets scheduled for attack in the ATO—to a second application, the Rapid Application of Air Power (RAAP). RAAP is used in the target development and weaponeering process, the output of which is the Target Nomination List (TNL) or a fully weaponeered set of targets. When completed, the TNL is transmitted to a third application called the Advanced Planning System (APS). APS is the core capability in CTAPS 5.0x. It is used for coordinating strike and support packages, air refueling planning, for refining the Time on Target (TOT) of strike aircraft, and for deconflicting many aspects of the ATO. The list of deconflicted strike and support package assignments, including TOTs, air refueling assignments and times, and other supporting information is called the Air Battle Plan (ABP). When the ABP is completed, it is transmitted from APS to CAFMS.[2] In CTAPS 5.0x, CAFMS is used to reformat the ABP and collate it with other supporting information. The ABP and these other data together form the ATO, which is then transmitted to the unit level.

Each application in Figure S.1 was originally developed as a stand-alone system and so each has its own independent databases. In each information transfer shown in Figure S.1, an entire database must be transferred if the transfer is done automatically. In a major conflict, planners will have to deal with large numbers of targets, threats, and air assets. In a major conflict, each such database transfer could take hours to complete. Furthermore, when changes are made to a database that is being used as input data in a CTAPS application (for example, the TNL in APS), all planning activity must cease until the database has been updated and is again "locked." Limited interoperability exists between the databases used by different CTAPS 5.0x applications.

Even though CTAPS 5.0x is a significant improvement over the system used during the Gulf War, it has limitations that could impede planners in a major conflict. If not addressed, these limitations will also severely restrict how the air campaign planning process can be restructured and may prevent solution of important planning problems encountered during the Gulf War.

CTAPS 6.0

Three new applications are planned for version 6.0 of CTAPS. The Force Level Execution (FLEX) system will be used to monitor ATO execution and to formulate changes to the published ATO. The Battlefield Situation Display (BSD) system will display graphically the situation awareness data and other information from CTAPS databases. The third application, the Air Campaign Planning Tool (ACPT), is an existing stand-alone computer-based decision aid that is used to select and prioritize targets and that can also be used to reduce the time needed for production of the Master Attack Plan (MAP).

---

[2]After Desert Storm, CAFMS software was ported to CTAPS computer work stations. CAFMS is an example of legacy software that has been reused in the CTAPS program.

Figure S.2 shows the key applications and data flows in the CTAPS 6.0 ATO production process. Previously, MAP production, an important part of the planning process, had to be done manually. In CTAPS 6.0, ACPT will be used to prioritize and select targets, and to reduce the time needed for MAP production.

CAFMS will no longer be used in the ATO production and dissemination process in CTAPS 6.0. ATO collation and dissemination will instead be performed by APS. Because CAFMS databases will no longer have to be updated during ATO production, the number of time consuming database transfers will be reduced, and the time needed for database coordination and management will be reduced as well.

ACPT will add important new capabilities to the CTAPS architecture; however, several system integration issues must be resolved so that the data flows indicated in the figure take place in a timely fashion. As a stand-alone application, ACPT can run at different security classification levels but is most effective at prioritizing targets if it operates at the highest possible classification level. Because a large staff, as well as potential coalition partners, may need to access CTAPS during operations, CTAPS is best run as a secret-level system. Consequently, a multilevel security (MLS) system is needed to integrate CTAPS and ACPT effectively. To limit the complexity and development risk for these MLS interfaces, they should be kept as simple as possible. In the body of this report, the details of several integration options are discussed.

FLEX and BSD promise to significantly increase the situation awareness and real-time command and control capabilities of the JFACC and his staff. However, to achieve this promise, several system and database integration issues must also be addressed. In particular, for BSD to display data from CTAPS databases in near real-time fashion, it will have to automatically access these databases, interpret the data, and copy the data requested in queries made by the JFACC or his staff. Below we discuss in more detail how CTAPS databases can be integrated together to address these and other issues.

The current CTAPS architecture is a product of the evolutionary acquisition approach used in its development. This approach has been called the "build a little, test a little" method of system development. Rapid prototypes were based on state-

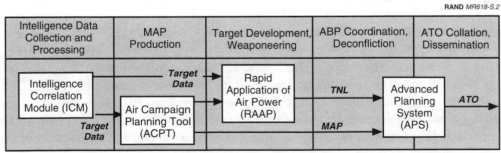

TNL: Target Nomination List, MAP: Master Attack Plan, ABP: Air Battle Plan, ATO: Air Tasking Order.

Figure S.2—Key Applications and Data Flows in the CTAPS 6.0 ATO Production Process

of-the-art commercial off-the-shelf (COTS) equipment and were taken out into the field and tested by operators. Legacy software from previous systems was ported to the same COTS equipment. This kept costs down and allowed quick fielding of an interim operational capability.

Despite the successful fielding of CTAPS 5.0x, attention should now be focused on better integrating existing CTAPS components—especially its databases—before, or at least along with, the integration of new applications such as FLEX and BSD into its already complex system architecture. Below we discuss these integration issues in more detail and describe the operational benefits that could be obtained from improved system integration in CTAPS.

REENGINEERING CTAPS AND SHORTENING THE ATO CYCLE

During Desert Storm, a relatively large number of last minute changes were made to the ATO in an effort to increase the responsiveness of high-value air-attack assets. This approach worked in the sense that some high-priority targets were attacked quickly, but it had several drawbacks. For example, mission crews frequently did not have enough time for mission planning.

If the ATO cycle can be shortened significantly, the responsiveness of the entire attack force—and not just a select subset of that force—can be increased. In this report, we indicate how the ATO planning cycle can be cut in half, to 24 hours, if the ATO production process is reengineered and certain key changes are made to CTAPS.

There are three elements to effectively reengineering CTAPS and shortening the ATO planning cycle:

- Provide automation support for MAP production.

- Reduce the number of target and order of battle (OB) databases in CTAPS, and automate target and OB database update processes.

- Perform target development and weaponeering in parallel with other ATO cycle processes, and schedule when target changes are transferred to MAP and ATO databases.

Automation Support for Master Attack Plan Production

Automating the MAP production process could significantly reduce the ATO planning cycle. In Desert Storm, about 11 hours were set aside for MAP production out of a total of 48 hours for the entire planning cycle. It should be possible to cut this time at least in half by using ACPT. ACPT also provides planners with important new capabilities that would be very difficult to duplicate by manual planning even with a large staff—the ability to quickly build several alternative notional MAPS and to compare their relative performance in meeting high-level strategic goals. By using ACPT, it may be possible to compare five or six notional MAPS, select one for refinement, and then refine it, all within four or five hours.

However, to use ACPT effectively with CTAPS, these two systems must be linked electronically. ACPT MAP data should be transferred automatically into CTAPS databases. Otherwise, significant time and manpower may be needed for manual data entry. Similarly, ICM OB and targeting data should be automatically transferred into ACPT. This does not necessarily mean ACPT and CTAPS have to use common target and OB databases. If ACPT operates at a higher classification level than CTAPS, as it probably should, an MLS messaging interface will be needed between the two systems. Targeting and OB information can be shared by transmitting precisely formatted messages through such an MLS interface.

Restructuring CTAPS Target and Order of Battle Databases

OB Databases. There are six separate OB databases in CTAPS 5.0x. They all have to be separately maintained and manually updated, which could be time consuming and difficult during combat operations. CTAPS applications with OB databases are probably best integrated by designating one application's database as the master OB database and giving that application, or its operator, control over OB data used in all CTAPS applications. ICM has been designated as the master OB database for CTAPS, although precisely what that means in CTAPS 5.0x is difficult to say. Additional steps are needed to ensure OB database integrity throughout the system. A systematic update process should be established to ensure that CTAPS OB databases are updated regularly and uniformly. Initially, these may have to be based on manual procedures.

In the long term, all CTAPS OB databases should be automatically linked and the number of OB databases reduced to alleviate the work load and time pressure on force-level planners. An integrated CTAPS database architecture is needed to ensure that OB databases remain synchronized. As the master OB database, ICM should automatically transmit OB updates throughout CTAPS. However, because randomly occurring OB updates could disrupt the planning process, these updates should probably be programmed to occur at preplanned times. That is, automatic OB updates should be scheduled. *If the ATO cycle is to be shortened, a unified and responsive CTAPS OB database architecture will be essential.*

Target Databases. At least four separate target databases are used in CTAPS 5.0x. Maintaining database integrity and ensuring all target databases are synchronized will also be challenging during combat operations. In the near term, one database should be designated as the master target database. If changes must be made manually, one duty station should be responsible for transmitting target changes to other database managers, and a clear set of procedures should be established for the target change process. *In the long term, the CTAPS database architecture should be modified so all CTAPS target databases are automatically linked to a single master target database, and if possible the number of target databases should be reduced.*

Common and Mirror Database Designs. A common CTAPS database design, in which mission applications would share one or more databases that would be automatically updated by CTAPS master databases, would allow CTAPS to be more readily adapted to different air campaign planning processes and would increase the flexibility with which planners and operators can use CTAPS. Mirror databases (databases that replicate a master database) would store data in a common format accessible by all mission applications. For example, a set of master and mirror target databases could be established, as illustrated in Figure S.3.

CTAPS 6.0 presents even more challenging database issues to systems integrators. It will be significantly more complex and will provide real-time access to the published ATO. The FLEX ATO database will have to remain current with real events. So will the BSD database. Both these applications may perform a large amount of real-time message processing, and their databases may have to be isolated from other mission applications to provide the level of responsiveness needed.

Parallel Target Development and Weaponeering and Scheduling Target Changes

The key to reducing ATO cycle time is to divide the subprocesses performed during the cycle into those that can be done in parallel or simultaneously and those that can be done only in series. Target development and weaponeering could be performed in parallel with MAP and ATO production.

Candidate targets would be developed, weaponeered, and then forwarded to combat planners and to ACPT. If new candidate targets were given a high enough priority, they would be grouped together in an ATO changes file. At preplanned times in the ATO planning cycle, all planning activity would stop, placing the ATO database into a static condition. Then the TNL in the APS ATO database could be changed and new high-priority targets added. A similar number of existing targets would be deleted from the TNL consistent with the apportionment guidance from higher authority. By developing and weaponeering candidate targets continuously during the cycle, targeteers could build up a library of fully weaponeered prioritized targets that could be quickly inserted at a suitable juncture into the ATO production process.

RAND MR010-S.3

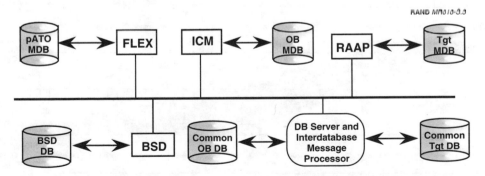

NOTES: DB: Database, MDB: Master Database, OB: Order of Battle, Tgt: Target, pATO: Published ATO.

Figure S.3—Notional CTAPS Air Operations Center Database Architecture

An important element of this change in the ATO production process is to introduce *a set schedule for making changes to the ATO* once actual ATO coordination and de-confliction is under way. This will help coordinate parallel planning activities and will probably be necessary when a large number of changes are made to the ATO.

ATO changes would be transferred between CTAPS databases on a prescheduled basis to reduce the number of potential disruptions experienced by ATO planners. However, to carry out these planning activities in parallel, CTAPS target databases have to be electronically linked so the transfer of target changes is made quickly and additional time delays are not introduced into the shortened ATO cycle.

Solving this database access problem presents technical challenges. If all CTAPS applications used the same target database, the number of database transactions may be too large to maintain overall system responsiveness. A common target database may have to be replicated or mirrored on a second server to provide responsive access for all CTAPS applications requiring such access.

Figure S.4 illustrates how the ATO cycle could be cut in half if ACPT was used to automate MAP production, target development and weaponeering were carried out in parallel with other ATO production subprocesses, and if CTAPS databases were linked appropriately. During a 24 hour planning cycle, planners and weaponeers would continuously identify, prioritize, develop, and weaponeer targets. When new high-priority targets are identified and weaponeered, they would be added to the ATO changes file, which would be inserted into the MAP or ATO production processes at prescheduled times in the planning cycle.

CTAPS AND ATO INTEROPERABILITY

The ATO will be disseminated to a large number of joint command and control centers and units by means of CTAPS and the Global Command and Control System

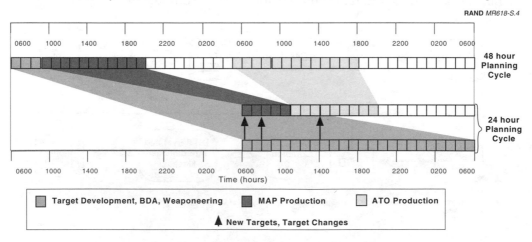

Figure S.4 —Notional Shortened ATO Cycle

(GCCS), as shown in Figure S.5. The figure indicates the large number of information system environments CTAPS and GCCS will have to be integrated into. The integration of CTAPS in this joint environment is a significant challenge.

The long-term approach to this problem was established when the Assistant Secretary of Defense for Command, Control, Communications, and Intelligence (C3I) mandated that future versions of CTAPS, as well as most other C4I information systems being developed by the services, will transition to the GCCS Common Operating Environment (COE)—when the GCCS COE becomes available. The GCCS COE is designed to provide a single common software environment that different service and joint command mission applications can run on, even if different types of computers are used. The GCCS COE will also have a number of basic support applications and communications capabilities built in to enable data and messages to be exchanged between different mission applications. The GCCS COE is being devel-

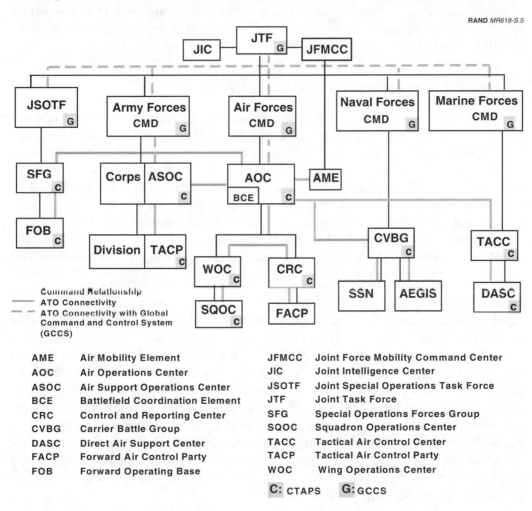

RAND MR618-S.5

Command Relationship
—— ATO Connectivity
- - - ATO Connectivity with Global Command and Control System (GCCS)

AME	Air Mobility Element	JFMCC	Joint Force Mobility Command Center
AOC	Air Operations Center	JIC	Joint Intelligence Center
ASOC	Air Support Operations Center	JSOTF	Joint Special Operations Task Force
BCE	Battlefield Coordination Element	JTF	Joint Task Force
CRC	Control and Reporting Center	SFG	Special Operations Forces Group
CVBG	Carrier Battle Group	SQOC	Squadron Operations Center
DASC	Direct Air Support Center	TACC	Tactical Air Control Center
FACP	Forward Air Control Party	TACP	Tactical Air Control Party
FOB	Forward Operating Base	WOC	Wing Operations Center

C: CTAPS   G: GCCS

Figure S.5 —Joint ATO Connectivity Supplied by CTAPS and GCCS

oped in parts by the services and will be integrated by the Defense Information Systems Agency (DISA).

In the near term, complex interfaces to other C4I systems will have to be developed to achieve the type of interoperability needed to *automatically* disseminate the ATO and to exchange situation awareness, force status, and intelligence information. Developing many complex interfaces to other C4I systems—many of which are undergoing rapid programmatic and technological change—could divert valuable resources from other important aspects of the program. CTAPS software should continue to be selectively ported to other C4I systems to provide the required joint ATO dissemination capability. CTAPS should also be able to electronically exchange raw information files that can then be manually processed by operators in a wide array of C4I systems. This type of low-level interoperability is easily achieved if Internet communication protocols continue to be used in CTAPS and other DoD C4I systems. More-advanced message processing features could be developed, but only after a broad set of appropriate information system standards and standard data elements have been agreed to by all the services.

Real-Time Joint Access to the ATO Database

With the introduction of FLEX in version 6.0 and the Air Support Operations Center (ASOC) automation program, the published ATO database will become a dynamic database for the JFACC and his staff, but also potentially for the much wider range of units and commanders as shown in Figure S.5. The challenge associated with the wider electronic dissemination of a dynamic *published* ATO is to provide an accurate and up-to-date copy of it, including all "last minute" ATO changes, to recipients who really need the additional changes. The units that may require real-time access to the published ATO during ATO execution are shown in Figure S.6. Today the airborne units shown in the figure do not have the capability to receive real-time ATO updates except by means of voice communications.

ATO databases at each unit will have to be synchronized with one another or will need to contain exactly the same information to avoid problems. Database mirroring is more difficult in this case because database updates will be transmitted using lower bandwidth communications media. During a real operation, it will be difficult to keep all the databases shown in the figure synchronized. They will frequently be inconsistent to some degree because of communications delays, processing bottlenecks, and because manual processing may be required at some units. Robust database links and adequate wide area network communications capacity will be needed to maintain synchronization of a dynamic ATO database. Difficult technical hurdles must be overcome and more robust computer network communications will be needed to provide this capability reliably. A few approaches are suggested in this report for dealing with this challenging problem.

RAND *MR618-S.6*

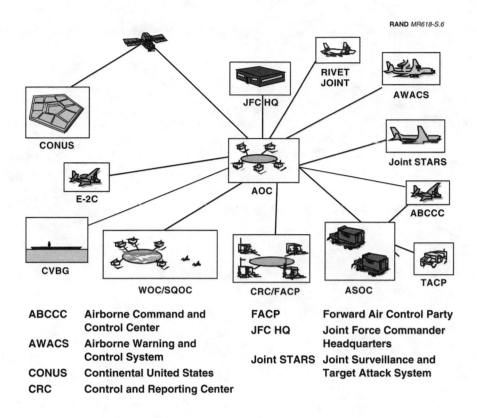

ABCCC	**Airborne Command and Control Center**
AWACS	**Airborne Warning and Control System**
CONUS	**Continental United States**
CRC	**Control and Reporting Center**

FACP	**Forward Air Control Party**
JFC HQ	**Joint Force Commander Headquarters**
Joint STARS	**Joint Surveillance and Target Attack System**

Figure S.6—Replicated Dynamic ATO Databases

ACKNOWLEDGMENTS

I would like to thank Major Jesse Citizen (Air Force Deputy Chief of Staff Plans and Operations, Directorate of Forces, Combat Integration Division), Tom Clark and Jim Papagni of Rome Laboratory, Lt. Col. William Craig (Air Combat Command, Special Management Office), and Lt. Col. David Waterstreet (Air Combat Command, Directorate of Combat Requirements) for their help in gathering information used in this report, and Col. Terry Oldham (Air Combat Command, Special Management Office) for providing RAND the opportunity to attend meetings of the Theater Battle Management "06 level" Advisory Group.

I also thank Leland Joe and Louis Moore of RAND for helpful discussions, and Patrice Roberts, Linda Quicker, and Sarah Young for their help in preparing this manuscript.

Finally, I wish to thank Myron Hura and Jack Craigie of RAND for their careful review of this report.

ABCCC	Airborne Command and Control Center
ABP	Air Battle Plan
ACEs	Airborne Command Elements
ACO	Airspace Coordination Order
ACPT	Air Campaign Planning Tool
ADM	Application Distribution Module
ADS	Airspace Deconfliction System
AFRN	Air Force Request Net
AFSAB	Air Force Scientific Advisory Board
AGCCS	Army Global Command and Control System
AI	Air Interdiction
AI	Artificial Intelligence
AIS	Automated Information System
ALCC	Airlift Command Center
ALLOREQ	Allocation Recommendation
AME	Air Mobility Element
AOC	Air Operations Center
AOOA	Air Operations Order and Apportionment
APPO	Application Portability Profile
APS	Advanced Planning System
ARPA	Advanced Research Projects Agency
ASDC3I	Assistant Secretary of Defense for C3I
ASOC	Air Support Operations Center
ATACCS	Advanced Tactical Air Command Central System
ATM	Air Tasking Message
ATO	Air Tasking Order
ATRM	ATO Transfer and Review Module
ATTD	Advanced Technology Transition Demonstration
AUTODIN	Automatic Data Interchange Network
AWACS	Airborne Warning and Control System
AWOP	Advanced Weaponeering Optimization Program

BASS	BCE Automated Support System
BCE	Battlefield Coordination Element
BDA	Bomb Damage Assessment
BE numbers	Basic Encyclopedia numbers
BSD	Battlefield Situation Display
C2	Command and control
C3I	Command, Control, Communications, and Intelligence
C4I	Command, Control, Communications, Computers, and Intelligence
CA	Counter Air
CAFMS	Computer Assisted Force Management System
CAFWSP	Combat Air Force Weather Support Program
CAP	Combat Air Patrol
CARS	Contingency Airborne Reconnaissance System
CAS	Close Air Support
CBUI	Character-based user interface
CGS	Common Ground Station
CID	Combat Intelligence Division
CIM	Corporate Information Management
CINCCENT	Commander in Chief, Central Command
CIO	Central Imagery Office
CIS	Combat Intelligence System
CISC	Complex Instruction Set Computer
CITA	Combat Intelligence Targeting on Arrival
CMS	Common Mapping System
COD	Combat Operations Division
COE	Common Operating Environment
COGs	Centers of gravity
CONUS	Continental United States
CORBA	Common Object Request Broker
COTS	Commercial off-the-shelf
CPD	Combat Plans Division
CPU	Central Processing Unit
CRC	Control and Reporting Center
CRP	Control and Reporting Party
CS	Constant Source
CTAPS	Contingency Theater Automated Planning Systems
CTEM	Conventional Targeting and Effectiveness Model
CTK	Communication Tool Kit
CTOs	Change Task Orders
CVBG	Carrier Battle Group

DASC	Direct Air Support Center
DCS	Data Communications System
DCT	Digital Communications Terminal
DIA	Defense Intelligence Agency
DISA	Defense Information Systems Agency
DISN	Defense Information Systems Network
DMPIs	Desired Mean Points of Impact
DMS	Defense Messaging System
DoD	Department of Defense
DREN	Defense Research and Engineering Networks
EC	Electronic combat
ENSCD	Enemy Situation Correlation Division
EOB	Electronic order of battle
FACP	Forward Air Control Party
FIPS	Federal Information Processing Standards
FLEX	Force Level Execution System
FOB	Forward Operating Base
FTP	File Transfer Protocol
GAT	Guidance Apportionment and Targeting
GB	Gigabyte
GCCS	Global Command and Control System
G&I	Guidance and Intentions
GKS	Graphical Kernal System
GOTS	Government off-the-shelf
GUI	Graphical user interface
HMI	Human Machine Interface
HVAA	High Value Airborne Assets
ICM	Intelligence Correlation Module
IDASS	Improved Direct Air Support System
IDB	Intelligence Database
IDM	Improved Data Modem
IFF	Identification friend or foe
IFR	Instrument Flight Rules
IMOM	Improved Many on Many
INTSUMS	Intelligence Summaries
IP	Internet Protocol
ISDN	Integrated Services Digital Network
IT	Information Technology
JDISS	Joint Intelligence Support System
JFACC	Joint Force Air Component Commander

JFC	Joint Force Commander
JFMCC	Joint Force Mobility Command Center
JIC	Joint Intelligence Center
JIPTL	Joint Integrated Prioritized Target List
JMCIS	Joint Maritime Command Information System
JMEM	Joint Munitions Effectiveness Manual
Joint STARS	Joint Surveillance and Target Attack System
JSIPS	Joint Service Imagery Processing System
JSOTF	Joint Special Operations Task Force
JTCB	Joint Target Coordination Board
JTF	Joint Task Force
JUDI	Joint Universal Data Interpreter
KBP	Knowledge base program
KTO	Kuwaiti Theater of Operations
LANs	Local area networks
LOS	Line-of-sight
MAP	Master Attack Plan
MAOC	Modular Air Operations Center
MCEB	Military Communications and Electronics Board
MC&G	Mapping, Charting, and Geodesy
MILSPEC	Military Specifications
MIME	Multipurpose Internet Mail Extension
MISREPs	Mission Reports
MISSI	Multilevel Information Systems Security Intiative
MLS	Multilevel security
MRC	Major Regional Conflict
MSBR	Manual System Backup to Recovery
NAFs	Numbered Air Forces
NALE	Naval Aviation Liaison Element
NAVFOR	Naval Force
NCA	National Command Authority
NIST	National Institute for Standards and Technology
OB	Order of battle
OCA	Offensive Counter Air
ODS	Operation Desert Storm
OLE	Object Linking and Embedding
OMG	Object Management Group
OORDBs	Object-Oriented Relational Databases
OS	Operating system
OSC	Air Force Intelligence Agency Operations Support Center

OSD	Office of the Secretary of Defense
OSI	Open Standards Interconnect
OT	Object Technology
PCs	Personal computers
PGM	Precision Guided Munition
POSIX	Portable operating system
RAAP	Rapid Application of Air Power
RDA	Remote Data Access
RDBMS	Relational database management systems
REM	Route Evaluation Model
RISC	Reduced Instruction Set Computer
RLIST	Routing List Management
SB	Sentinel Byte
SCI	Special Compartmented Intelligence
SCRAM	System Configuration and Reconfiguration Automation Module
SDM	System Diagnostic Module
SEAD	Suppression of Enemy Air Defenses
SFG	Special Forces Group
SFG	SOF Group
SFOB	SOF Forward Operating Base
SIGS	Secondary Imagery Graphics System
SMA	System Message Alert
SOF	Special Operations Forces
SONET	Synchronous Optical Network
SPINS	Special Instructions
SQL	Structured query language
SQOC	Squadron Operations Center
SSM	System Security Module
STT	Strategy-to-tasks
TACC	Tactical Air Control Center
TACP	Tactical Air Control Party
TACPs	Tactical Air Control Parties
TACS	Tactical Air Control System
TADIL	Tactical Digital Interface Link
TAFIM	Technical Architecture for Information Management
TARBULs	Target Bulletins
TBMCS	Theater Battle Management Core System
TCP	Transmission Control Protocol
TCP/IP	Transmission Control Protocol/Internet Protocol
TD&W	Target development and weaponeering

TIBS	Tactical Information Broadcast Service
TMD	Theater missile defense
TNL	Target Nomination List
TOTs	Times on Target
TPN	Tactical Pocket Network
TPWs	Target Planning Worksheets
TRA	Technical Reference Architecture
TRAP	Tactical Receive Equipment and Related Applications
TPW	Target Planning Worksheet
UFLink	Unit-Force-Level Link
UGDF	Army Uniform Guided Data Field
UI	User interface
USCINCLANT	Commander and Chief U.S. Atlantic Forces
USCINCPAC	Commander and Chief U.S. Pacific Forces
USI	User interface
USMTF	U.S. Message Text Format
VFR	Visual Flight Rules
WAN	Wide area network
WOC	Wing operations center
WCCS	Wing Command and Control System
WWMCCS	World Wide Military Command and Control System
XIDB	Extended intelligence database

Chapter One

INTRODUCTION

Operation Desert Storm (ODS) air operations were some of the most complex in the history of warfare. At certain points during the war, over 3,000 sorties a day were flown with a third of these being combat sorties. Despite the size of the operation and the fact that aircraft from many coalition countries flew under Joint Force Air Component Commander (JFACC) direction, the United States and its allies suffered no fratricide in air-to-air combat. The elimination of air-to-air fratricide was due to several factors, one of which was the tightly knit air campaign plan that all fixed wing aircraft followed each day in the theater of operations.[1]

During the Gulf War the JFACC was confronted with trying to achieve two conflicting goals. On the one hand, he wanted to employ air power in the most effective and well orchestrated way possible. This was done by having his staff produce the Air Tasking Order (ATO) in a well-ordered, deliberate planning process by disseminating the ATO in a timely fashion to the unit level and ensuring that air operations did not deviate too far from the ATO. On the other hand, the JFACC also needed to change target priorities and attack new high-priority targets as quickly as possible. But if too many changes were made to the ATO in too short of a time, the planning process can be severely impacted leading potentially to chaos and a dramatic reduction in the effectiveness of the overall air campaign. A balance must be struck between these two goals to maximize the application of air power in large air campaigns.

During ODS it was hard to balance these goals because of the difficulties encountered in producing and disseminating the ATO. Although an automation aid was used to produce and disseminate the ATO, the Computer Aided Force Management System (CAFMS), it had numerous drawbacks. It was based on obsolete computer hardware and could only display ATO data as difficult to interpret text. Only with tremendous difficulty was it modified so that it could support the large sortie rate of coalition air forces. Because of the large number of sorties in each ATO and computer system limitations, it was very difficult to adjust the ATO properly when new high-priority targets or target changes were added in the latter stages of the ATO production process. And even though planners had almost 40 hours to plan air attacks

---

[1]Mission plans for nearly all aircraft which operated over land in the area of operations were required to be in the Air Tasking Order (ATO). Navy and Marine Corps aircraft which operated over water and which provided fleet defense or other naval support missions were not in the ATO.

and prepare the ATO, they struggled throughout the war to produce a fully coordinated and deconflicted ATO within that time period.

Significant difficulties were encountered in obtaining, deconflicting, and properly incorporating order of battle and bomb damage assessment information into the ATO planning process. There were several reasons for these problems, but one important factor was the lack of automation support for storing and processing this type of information and for other vital planning activities. Finally, many units received the ATO late or with tremendous difficulty. Again there were several reasons for the latter problem, but a primary culprit was the CAFMS communications system.

Before ODS numerous programs were under way in the Air Force and DoD to address many of these deficiencies. In the last few years many of these programs were combined into an umbrella program in an effort to better coordinate these development activities. The purpose of this umbrella program is to develop a state-of-the-art force-level Command, Control, Communications, Computers and Intelligence (C4I) system, called until recently the Contingency Theater Automated Planning System (CTAPS), that can support all phases of ATO production, dissemination, and execution monitoring.[2]

In this report we examine CTAPS and the air campaign planning process and propose modifications to both which will help remedy these deficiencies and dramatically increase the responsiveness of U.S. air forces in large conflicts. We examine the functionality of CTAPS and the structure of the air campaign planning process in detail. Below we provide a top-level description of CTAPS and an outline for the remainder of this report.

CTAPS 5.0x

CTAPS is a complex automated support system that runs on a large networked set of computer work stations. Because of its ancestry and the evolutionary acquisition approach used in its development, CTAPS is a complex combination of applications that have been modified to run together with minimal interference in the same distributed client-server computing environment. The four key applications used in the ATO production process in version 5.0x of CTAPS and the key data flows in the process are shown in Figure 1.1. There are many more applications in CTAPS 5.0x; however for simplicity we have included only the major ones in the figure (see Figures 5.1 and 7.1 for more detailed illustrations).

The ATO planning cycle starts with the collection of intelligence data needed for air campaign planning. The Intelligence Correlation Module (ICM) is used to correlate order of battle, situation awareness and target intelligence information received from external sources. In many CTAPS configurations ICM supplies target data and the set of targets scheduled for attack in the ATO to a second application, the Rapid

---

[2]CTAPS has recently been renamed the Theater Battle Management Core System (TBMCS). Since much of the analysis in this report concerns specific software versions of CTAPS, for simplicity we refer to future versions of TBMCS as CTAPS.

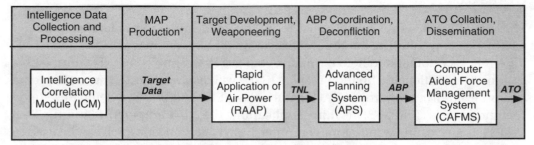

TNL: Target Nomination List, MAP: Master Attack Plan, ABP: Air Battle Plan, ATO: Air Tasking Order.
*Manual Process if only CTAPS 5.0x is available.

Figure 1.1—Key Applications and Data Flows in the CTAPS 5.0x ATO Production Process

Application of Air Power (RAAP). RAAP is used in the target development and weaponeering process, the output of which is the Target Nomination List (TNL) or a fully weaponeered set of targets. When completed, the TNL is transmitted to a third application called the Advanced Planning System (APS). APS is the core capability in CTAPS 5.0x. It is used for coordinating strike and support packages, air refueling planning, for refining the Time on Target (TOT) of strike aircraft, and for deconflicting many aspects of the ATO. The list of deconflicted strike and support package assignments, including TOTs, air refueling assignments and times, and other supporting information is called the Air Battle Plan (ABP). When the ABP is completed it is transmitted from APS to CAFMS.[3] In CTAPS 5.0x CAFMS is used to reformat the ABP and collate it with other supporting information. The ABP and these other data together form the ATO, which is then transmitted to the unit level.

Each application in Figure 1.1 was originally developed as a stand-alone system, and so each has its own stand-alone databases (for example in CTAPS 5.0x there are six separate order of battle (OB) databases). In each information transfer shown in Figure 1.1 an entire database must be transferred if the transfer is done automatically. In a Major Regional Conflict (MRC) planners will have to deal with large numbers of targets, threats, and air assets. In an MRC each such database transfer could take hours to complete. Furthermore, when changes are made to a database that is being used as an input in a CTAPS application (for example, the TNL in APS), all planning activity must cease until the database has been updated and is again "locked." Consequently, limited interoperability exists between the databases used by different CTAPS 5.0x applications. These limitations restrict how the air campaign planning process can be structured and changed.

---

[3]After Desert Storm CAFMS software was ported to CTAPS computer work stations. It is an example of legacy software that has been reused in the CTAPS program.

CTAPS 6.0

Three new applications are planned to be included in the next major upgrade of CTAPS (version 6.0). The Force Level Execution (FLEX) system will be used to monitor ATO execution and to help formulate changes to the published ATO during execution. The Battlefield Situation Display (BSD) will display situation awareness data and other information from CTAPS databases. FLEX and BSD are being developed specifically for CTAPS. The third application, the Air Campaign Planning Tool (ACPT), is an existing stand-alone computer-based decision aid which can be used to select and prioritize targets based on high-level campaign objectives and which can also be used to reduce the time needed for production of the Master Attack Plan (MAP).

Figure 1.2 shows the key applications and data flows in the CTAPS 6.0 ATO production process. Previously, MAP production, an important part of the planning process, had to be done manually. In CTAPS 6.0 ACPT will be used to prioritize and select targets, and to reduce the time needed for MAP production.

As indicated in Figure 1.2, CAFMS will no longer be used in the ATO production and dissemination process in CTAPS 6.0. ATO collation and dissemination will instead be performed by APS. Because CAFMS databases will no longer have to be updated during ATO production, one less-time-consuming database transfer or update process will have to be supported, and the time needed for database coordination and management will be reduced.

ACPT will add important badly needed capabilities to the CTAPS architecture; however, there are several system integration issues that must be resolved to ensure that the data flows indicated in the figure take place in a timely fashion. On the one hand, as a stand-alone application ACPT can run at different security classification levels, but is most effective at prioritizing targets if it operates at the highest possible classification level. On the other hand, CTAPS operates at the secret classification level. Because a large staff, as well as potential coalition partners, may have access to CTAPS during operations, CTAPS is best run as a secret-level system. Consequently, a multilevel security (MLS) system is needed to integrate CTAPS and ACPT

RAND *MR618-1.2*

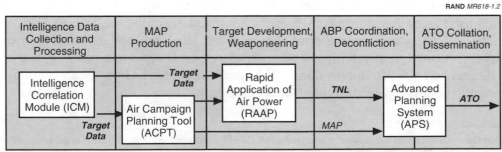

TNL: Target Nomination List, MAP: Master Attack Plan, ABP: Air Battle Plan, ATO: Air Tasking Order.

Figure 1.2—Key Applications and Data Flows in the CTAPS 6.0 ATO Production Process

effectively. In the body of this report the details of several integration options are discussed. Here we only point out that in order to limit the complexity and development risk for these MLS interfaces, they should be kept as simple as possible.

Not shown in the figure are FLEX and BSD, the two new applications that will be used to monitor and adjust the "current" or published ATO that is being executed. They promise to significantly increase the situation awareness and real-time command and control capabilities of the JFACC and his staff. However, in order for this promise to be realized several system and database integration issues must also be addressed. In particular, for BSD to display data from CTAPS databases in near real-time fashion, it will have to be capable of automatically accessing these databases, interpreting the data, and copying the data it needs in response to the queries from the JFACC or his staff.

OBJECTIVES

The current CTAPS architecture has resulted from the evolutionary acquisition approach used in its development. This approach has been called the "build a little, test a little" method of system development. Rapid prototypes were based on state-of-the-art commercial off-the-shelf (COTS) equipment and were taken out into the field and tested by operators. Legacy software from previous systems was ported to the same COTS equipment. This kept costs down and allowed quick fielding of an interim operational capability.

Rapid technological change in COTS-based information systems, budget cuts, and new Office of the Secretary of Defense (OSD) policy direction to terminate C4I systems based on legacy technologies have led to an environment of rapid programmatic change for C4I systems, and to a pressure to combine or eliminate similar programs. These trends have reduced the duplication of effort between C4I system programs but have presented new challenges to the Air Force. CTAPS is one of the few Air Force C4I programs whose scope has increased significantly in the past few years. It is being deployed to units of all the services, which means that CTAPS software or hardware will be inserted into a variety of different computer environments. The new expanded scope of the program may make it more difficult to integrate CTAPS and related systems into a single seamless architecture. One objective of this report is to identify system development options that can meet these challenges and result in a single unified C4I system architecture for air campaign planning and execution.

The primary objectives of the report are to examine the air campaign planning process, to seek lessons from past experience on how to improve the process, and to identify new procedures and modifications to CTAPS that can significantly reduce the time needed to produce the ATO and hence to increase the effectiveness of U.S. combat air forces. One caveat to this analysis should be mentioned. The timely delivery of intelligence information is also necessary to carry out the air campaign planning process. However, the design, tasking, and linkages of external intelligence collection and processing systems are beyond the scope of the current investigation and will not be considered in this report.

REPORT OUTLINE

In Chapters Two and Three we review the air campaign planning process. In Chapter Two we examine key elements of the process, the organizations, information flows, timelines, and command decision points involved. In Chapter Three we examine how the process was carried out during the Gulf War, and how it has been carried out in recent exercises. We identify ways to improve the air campaign planning process based upon lessons learned from the Gulf War and recent exercises.

In Chapter Four we examine in detail the current CTAPS 5.0x architecture. We analyze the underlying software architecture and identify its key components, including the communications programs, relational database management systems, and other supporting applications that form the foundation of the current system. Next we examine the mission applications and the functions they perform in the air campaign planning process.

The ACPT and related intelligence support automation tools are examined in Chapter Five. The design and capabilities of ACPT are examined in detail, as are the functions, relationships, and connectivity of the intelligence support tools used in the Air Operations Center. These intelligence support systems must work together with and within CTAPS for the entire air campaign planning process to be carried out effectively.

In Chapter Six we examine in detail the planned CTAPS 6.0 architecture. Details of the underlying software architecture were not yet available when this research was done, so the underlying software architecture is not described. However, we examine in detail the new mission applications that are planned to be integrated into CTAPS, and the functions they will perform in the air campaign planning process. We also analyze several options for integrating ACPT into CTAPS.

In Chapter Seven we examine in detail how the current ATO production process is implemented in the current CTAPS 5.0x architecture and the key information flows in the process. We identify the impediments to improving the process that are due to current CTAPS limitations. Next, we examine how the process could be implemented in CTAPS 6.0 if certain changes were made to the underlying database architecture of CTAPS and if ACPT is integrated effectively into the system.

Finally, in Chapter Eight we apply insights gained in earlier chapters and examine, from a process reengineering perspective, issues associated with the future evolution of the CTAPS architecture. We show that if several key issues are addressed and certain new capabilities added to CTAPS, the current ATO planning cycle time of 48 hours can be cut in half. However, in order to achieve this goal, attention must be focused on better integrating the components of CTAPS—especially its databases— that already are a part of the system.

The majority of the data collection for this study took place in 1994. However, because publication was delayed, information pertaining to CTAPS developments was gathered until November 1995 and was used in the final version of this report.

Chapter Two

AIR CAMPAIGN PLANNING IN THEORY

Much is involved in air campaign planning. The JFACC in particular has to be concerned with devising an overall air campaign that satisfies the military campaign objectives of the JFC and the national security objectives of the president and Joint Chiefs of Staff. Such a campaign plan is usually divided into a series of phases. The first phase may have as its objectives the attainment of air superiority or the destruction of the enemy integrated air defense network. Later phases may focus on destroying key target sets, such as enemy weapons of mass destruction. Although important, high-level strategic aspects of air campaign planning are not the focus of this report. These aspects of air campaign planning, and in particular JFACC responsibilities and related doctrine, are treated in detail in the Air Force *JFACC Primer* and elsewhere.[1]

Instead we focus on the "mechanics" of the air campaign planning process, the command and control (C2) and coordination mechanisms used by the JFACC to direct air forces under his command, and the sequence of steps necessary to coordinate air forces operating under different command relationships. In this chapter we briefly review the essential elements of the air campaign planning process:

- Organization

- The Air Tasking Order

- Required information flows

- Timelines

- Key command decision points internal to the process.

ORGANIZATION

A number of different organizations can be involved in air campaign planning, depending upon the size or type of conflict, and the composition of air forces involved. In the case of a single "surgical" strike, like the attack against Libya, planning may take place at high-level organizations in Washington and no forward-based planning organizations may exist. However, in an emerging crisis, while air and naval forces

---

[1]USAF, 1994b. Aspects of air campaign planning are also discussed in Thaler and Shlapak, 1995.

are deploying to the region, planning may take place at several different locations: at sea in carrier battle groups, on airborne command posts, and at Continental United States (CONUS)–based support organizations. In this latter case, the services face a difficult challenge in coordinating dispersed forces and the activities of distant command and control organizations, some of which may have limited planning or information processing capabilities. One should consider a broad set of scenarios, including those above, to determine the full set of organizational needs for air campaign planning. Such an analysis is beyond the scope of the present investigation however. In this report we focus on the air campaign planning process and the forward deployed air campaign planning and execution management organizations required for large air campaigns.

At the force level the Air Force organization responsible for air campaign planning is the Air Operations Center (AOC). The organization of the AOC, or equivalent organizations, has differed in Numbered Air Forces (NAFs) and in different theaters, due to theater-unique aspects of command and control organization.

The AOC is typically composed of the four divisions shown in Figure 2.1, and the two liaison groups indicated by the central box in the figure. The Combat Plans Division (CPD) is responsible for building the MAP and the ATO. The Combat Intelligence Division (CID) is responsible for target development, intelligence projections, and for weaponeering (the latter function may also be performed in the other divisions of the AOC). The Combat Operations Division (COD) monitors and controls ATO execution and is responsible for maintaining the ATO after it has been completed by the CPD. The Enemy Situation Correlation Division (ENSCD) is responsible for current intelligence and for providing the COD with target updates, BDA, and new target information.

Depending upon the composition of the air forces under JFACC command, many other organizations may be involved in the air campaign planning process. Figure 2.2 illustrates the connectivity between the AOC, joint headquarters, component commands, and other units that contribute information to the air campaign planning process, or that need to receive the ATO. The figure illustrates both command relationships and ATO connectivity between these organizations. Those units which

RAND *MR618-2.1*

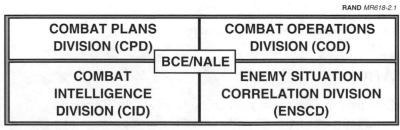

BCE: Battlefield Coordination Element.
NALE: Naval Aviation Liaison Element.

SOURCE: Cohen et al., 1993, Figure 15, p. 141.

Figure 2.1—Air Operation Center Divisions

have or will soon acquire CTAPS equipment for ATO dissemination are indicated by a small "c" in the lower righthand corner of each box. CTAPS ATO dissemination software has been named the joint standard for ATO dissemination. This component of CTAPS is planned for integration into the Global Command and Control System (GCCS). GCCS will be deployed at the JTF and component commander levels and will be able to access the CTAPS ATO database. These ATO connectivity links are illustrated in the figure by dashed gray lines, and units that will be equipped with GCCS are indicated by a small "G" in the lower righthand corner of each box.

For simplicity, what is not shown in Figure 2.2 are all potential sources of targeting and general intelligence information that may be used for air campaign planning. These additional sources of information could be a theater Joint Intelligence Center (JIC), a Unified Command JIC, or CONUS-based intelligence support organizations, such as the Defense Intelligence Agency (DIA) or the Air Force Intelligence Agency Operations Support Center (OSC). Many such remote information sources contributed to the targeting process during ODS.

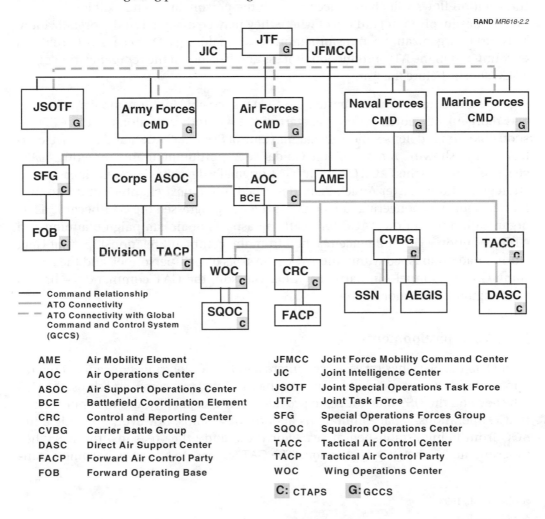

RAND *MR618-2.2*

AME	Air Mobility Element	JFMCC	Joint Force Mobility Command Center
AOC	Air Operations Center	JIC	Joint Intelligence Center
ASOC	Air Support Operations Center	JSOTF	Joint Special Operations Task Force
BCE	Battlefield Coordination Element	JTF	Joint Task Force
CRC	Control and Reporting Center	SFG	Special Operations Forces Group
CVBG	Carrier Battle Group	SQOC	Squadron Operations Center
DASC	Direct Air Support Center	TACC	Tactical Air Control Center
FACP	Forward Air Control Party	TACP	Tactical Air Control Party
FOB	Forward Operating Base	WOC	Wing Operations Center

C: CTAPS   **G:** GCCS

Figure 2.2—AOC Connectivity to Joint and Component Command Units

ODS Organization

The force-level organization responsible for air campaign planning and control in ODS was called the Tactical Air Control Center (TACC) which has since been retitled the AOC. During Desert Shield a TACC organization was established which followed the organizational paradigm depicted in Figure 2.1. The TACC was a large and complex organization. Total TACC personnel numbered over 2,000 if all intelligence and support personnel are included. For a number of reasons the TACC was reorganized shortly before the war started. As the war proceeded and as problems cropped up in the ATO preparation process, the structure and relationship of the Combat Plans and Combat Operations Divisions changed.[2]

Figure 2.3 is a theoretical depiction of an AOC organization. Organizational structures are a function of the individuals involved as much as they are of any organizational plan. If a new group of individuals is inserted into an existing organization, the functional organizational structure will inevitably change in unanticipated ways. Information flows will change according to the personalities and skill levels of the individuals involved, regardless of where they may be assigned in the organization. This sort of organizational transformation occurred during ODS because of problems encountered in the ATO production process and because of the reorganization of the TACC shortly before the war.

Figure 2.3 illustrates how the part of the CPD responsible for air campaign planning was reorganized shortly before the start of the war (other portions of the CPD responsible for air defense, the Tactical Air Control System (TACS), and airspace control are not shown). A new cell, the Guidance Apportionment and Targeting (GAT) cell, was added to the TACC CPD. The GAT originally developed overall air campaign strategy and the Master Attack Plan (described in the next chapter), and also performed some weaponeering functions. As the war progressed the GAT became a key organization within the TACC for nearly all aspects of air campaign planning and ATO execution. This influence resulted from the command relationships that Gen. Horner had established within the TACC, the skills of GAT personnel, and the access some GAT personnel—in particular Gen. Glosson, the GAT commander—had to Washington information sources.

Joint AOC Developments

After ODS the Navy has taken a keen interest in JFACC doctrine, concepts of operation, and in AOC organization.[3] The Navy has outfitted two command ships, the USS Whitney and the USS Blue Ridge, to provide an AOC capability and to support Navy JFACC operations. Space limitations aboard these ships, however, preclude a full AOC from being established at sea. For these and other reasons, the Navy has experimented with AOC organization. The GAT appears to have not become a stan-

---

[2]Cohen et al., 1993.
[3]USAF, 1993.

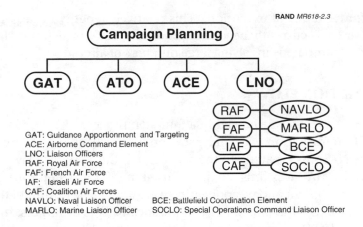

RAND *MR618-2.3*

GAT: Guidance Apportionmont and Targeting
ACE: Airborne Command Element
LNO: Liaison Officers
RAF: Royal Air Force
FAF: French Air Force
IAF: Israeli Air Force
CAF: Coalition Air Forces
NAVLO: Naval Liaison Officer
MARLO: Marine Liaison Officer

BCE: Battlefield Coordination Element
SOCLO: Special Operations Command Liaison Officer

SOURCE: Cohen et al., 1993, Figure 28, p. 198.

Figure 2.3—Desert Storm Air Campaign Planning Organization

dard part of Air Force AOC organization, but Navy implementations of the JFACC concept in recent exercises (in particular in recent Ocean Venture and Tempo Brave exercises) have included a GAT cell in the AOC (the GAT is sometimes called the Joint Targeting Cell in Navy publications).

The Navy and Marine Corps have also focused on AOC organization because of a concern over how naval air assets were viewed and employed by the JFACC during ODS. Some naval personnel have expressed the concern that there was insufficient naval representation within the TACC. An expanded Naval Aviation Liaison Element (NALE) has been proposed as a way to improve communications between the JFACC and naval component commands, and to enhance the JFACC's knowledge of naval assets and procedures.[4] With the addition of CTAPS to Navy and Marine Corps air operations C2 units and improved communications between the AOC and these units, the flow of information between them should be much better than that provided during ODS.

AIR TASKING ORDER

Orchestration of all the diverse elements of air power is accomplished by means of the ATO. It is the planning structure that provides detailed direction to air forces. It enables the JFACC to synchronize air attacks for maximum effect on the enemy, ensure the efficient use of air assets, and reduce fratricide during execution.

The ATO is a schedule that contains all information necessary to direct and coordinate air activity in the theater of air operations. Because the air campaign planning process depends on so many different pieces of information, it is a complex and time consuming process to produce the ATO. The current ATO cycle consists of two days

---

[4]Here the term naval implies both Marine Corps and Navy assets or personnel.

of planning for each day of execution. This relatively lengthy cycle, which is illustrated in Figure 2.4, is currently needed to produce a "self-consistent," "flyable" ATO, especially for air campaigns involving large numbers of aircraft.

ATO INFORMATION FLOWS

During ODS, the ATO was produced in a four step process. In the first step intelligence and BDA information was analyzed, and a prioritized set of targets to be attacked on Day N+2 was created. In the second step the MAP for Day N+2 was produced. When the MAP was finished, it was used in the third step during which detailed target development and weaponeering were done. The output of the third step were Target Planning Worksheets (TPWs) for specific sets of targets. In the fourth step, the MAP and TPWs were used to produce the ABP. Any necessary coordinating information, such as air corridors, call signs, identification friend or foe (IFF) codes, or communications frequency assignments for aircraft, was also produced and deconflicted in the fourth step and combined into the ABP, the Airspace Coordination Order (ACO) or the Special Instructions (SPINS). Finally, in the last part of step four, the ABP was combined with the ACO and SPINS to form the ATO.

The information needed to put the MAP, TPWs, and ATO together is illustrated in Figure 2.5. In the first step a set of prioritized targets was selected. This set was based on Joint Force Commander (JFC) general guidance, which was delivered to the JFACC in the Air Operations Order and Apportionment (AOOA) message, a list of prioritized targets established by the Joint Target Coordination Board (JTCB), and on JFACC guidance regarding strategic targeting objectives. In the AOOA, the JFC also apportioned air assets to different mission areas: Close Air Support (CAS), Air Interdiction (AI), and Counter Air (CA). Prospective new targets were also provided to the JFACC and MAP planners by intelligence analysts working in the TACC.

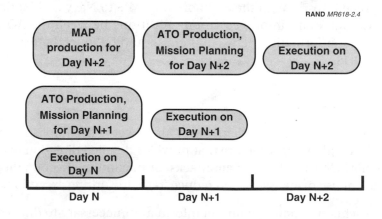

RAND MR618-2.4

Figure 2.4—Current ATO Planning and Execution Timelines

RAND *MR618-2.5*

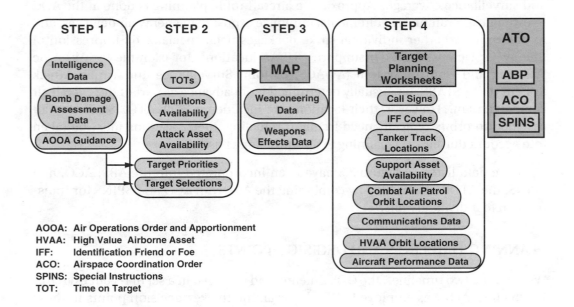

Figure 2.5—ATO Information Flow

In the second step the MAP was produced. It is essentially a list of aircraft assigned to specific strike packages which are in turn assigned to specific targets. Basic Encyclopedia Numbers (BE numbers) and TOTs were included for each strategic target listed in the MAP. Some support aircraft, such as Offensive Counter Air (OCA) escorts or jamming aircraft were usually listed in the MAP worksheets.

MAP planners also required the other types of information listed in Figure 2.5, such as which attack assets were available for day N from Air Force wings or carrier battle groups. They also needed data on munitions availability (numbers available and their location). Using this information, the MAP was constructed by matching available attack assets to targets.

In the third step in the process, weaponeering and detailed target development were performed at the force level at the TACC. The output of this step were TPWs.

The fourth step of ATO production is also illustrated in Figure 2.5. The force packages and TOTs defined in the MAP serve as the starting point for the ABP. Communications channels, IFF codes, call signs, and tankers are assigned to aircraft. In addition, weaponeering data—munition assignments and aim points—are included for strategic targets. The most complex part of ABP development is the coordination needed to ensure that strike packages form at the right place and time, without wasting fuel or without unduly exposing friendly aircraft to enemy threats.

An equally important part of the planning process is the allocation of support aircraft to strike packages, High Value Airborne Assets (HVAA), and Combat Air Patrol (CAP) aircraft. Tankers and fuel must be allocated to aircraft in a way which preserves the desired time phasing of offensive air strikes and which sustains airborne air defense and surveillance coverage. Approximate aircraft route planning is done at the AOC to estimate aircraft fuel requirements, minimize exposure to enemy air defenses, and to determine whether individual strike packages need specialized Suppression of Enemy Air Defenses (SEAD) support.[5] HVAA locations, for example for Airborne Warning and Control System (AWACS) or Joint Surveillance and Target Attack System (Joint STARS), are usually determined far in advance as part of the overall air campaign plan. However, their locations are included in each ATO. Since aircraft assigned to orbits typically need frequent refueling, their locations must be taken into account during ATO planning to optimize tanker assets.

After the data listed in Figure 2.5 have been incorporated into the ABP, ACO, and SPINS, the ATO is assembled by combining the ABP with ACO and SPINS for transmission to individual units.[6]

PLANNING TIMELINES AND DECISION POINTS

We examine two timelines, the ODS timeline and one used in a set of recent exercises in which CTAPS was employed. We also examine the key decision points in these processes.

ODS Timeline

The ODS ATO planning timeline was 48 hours long and was composed of the five stages illustrated in Figure 2.6. The first four stages of the timeline correspond to the four steps in the ATO planning cycle described above. The last stage in the timeline shown in the figure corresponds to the time allocated for ATO transmission to the units and for unit-level mission planning activities.

The timeline shown in Figure 2.6 is the desired timeline ODS planners used to structure the ATO production process. However, problems were encountered in ATO production during war, especially in the first few weeks. Initially, the timeline shown was not strictly adhered to. However, as the war progressed ODS planners were eventually able to meet the timeline shown. In the next chapter we will examine the problems encountered during ODS in some detail.

In the first stage, targets were prioritized and selected by the Commander in Chief Central Command (CINCCENT) and the GAT. The first stage of the cycle was approximately three hours long.

The second stage, MAP production, lasted 12 hours and was performed by the GAT.

---

[5]More precise route planning and target development are done at the unit level as part of the mission planning process.

[6]The ACO is a map of the airspace in the theater of operations which defines missile and air engagement zone boundaries, ingress and egress corridors, and other airspace control zones. The SPINS contain changes in operating procedures or the rules of engagement.

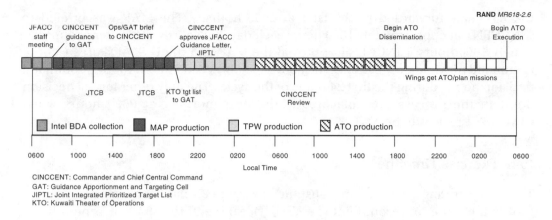

Figure 2.6—ATO Cycle for Operation Desert Storm

In the third stage, TPWs were produced. This process took about eight and a half hours to complete. The TPWs produced by GAT personnel listed preferred weaponeering options. Very limited automated support for weaponeering was available at the TACC during ODS. The TPWs by themselves contained most information needed to put together an ATO, including such things as call signs, tanker assignments, and escort assignments. However, TPW coordination was preliminary in nature and required further refinement.

Further refinement and coordination were supplied during the fourth stage of ATO construction. This stage was supposed to be completed in 14 hours. Most of this time and frequently more was needed to construct the ABP because of its complexity. At the conclusion of this stage, the ABP, the ACO, and SPINS were merged to form the ATO.

In the fifth and final stage, the ATO was transmitted to the units where mission planning was carried out. As shown in the figure, the wings were supposed to have about 11 hours to receive the ATO and to perform mission planning.

ODS Decision Points

Figure 2.6 also shows some key command decision points in the ODS ATO planning cycle. Most formal decisions were made during the first 24 hours of the planning process. First, CINCCENT supplied initial apportionment and high-level targeting guidance to the GAT at 0800 hours. The JTCB met twice daily, at 1200 and at 1700 hours. At these meetings component command representatives presented their own prioritized target lists for the Kuwaiti Theater of Operations (KTO). The JTCB prioritized these requests and merged them into what was called the Combined, Joint, Prioritized, and Integrated Target List (and what is now usually called the Joint Integrated Prioritized Target List (JIPTL)). The formal JTCB-approved target list for the KTO was forwarded to the GAT at 2000 hours.[7] The MAP was briefed to CINCCENT at approximately 1800 hours each day and shortly thereafter at approximately 1900 hours, CINCCENT approved the MAP and the U.S. Air Force Central Command Air Guidance Letter.[8] The above meetings constituted the key command decision points during the first 24 hours of the cycle. The only other formal decision point in the process took place early the next morning at 0800 hours when CINCCENT reviewed the ATO.

Joint Exercise Timelines

The ATO planning cycle timeline illustrated in Figure 2.7 has been used as a training goal in a number of recent joint exercises (Ocean Venture 1993 and Tempo Brave 94-1). This timeline resembles the one used in ODS but differs from it in some important ways.

Each ATO cycle started with three hours for preliminary target development. The second stage for MAP production was nominally only 7 hours long in these exercises.[9] In contrast, almost as much time was used for ATO production in both cases: at least 14 hours during ODS and at least 12 hours during the exercises.[10] The same amount of time was budgeted in both cases for ATO dissemination and unit-level mission planning.

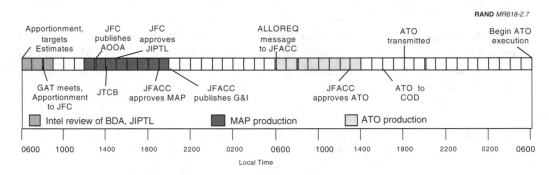

SOURCE: Commander in Chief U.S. Pacific Forces, n.d.

Figure 2.7—ATO Cycle for Exercises Ocean Venture 1993 and Tempo Brave 94-1

---

[7]Additional targeting requests and changes occurred during the course of the ATO execution day for air support or for interdiction in the KTO.

[8]This message is now called the JFACC Guidance and Intentions (G&I) Message or the Air Tasking Message (ATM). A common terminology may not exist in all NAFs and theaters.

[9]This may have been the case in the exercises because of the relatively small size of the exercise ATOs, because less high-level strategic planning was done, or because less-detailed target development and weaponeering were performed. A manual MAP process was used in both ODS and the exercises.

[10]In exercise Tempo Brave 94-1 the ATO was sometimes disseminated later than 1800 local time because of technical difficulties and training difficulties.

Note that a separate TPW activity is not shown in Figure 2.7. Targeting was done during the exercises at the AOC in parallel with MAP development. Target development and weaponeering was done by the CID using CTAPS prior to ATO development. Weapons selections were included in the target list when it was forwarded using CTAPS to ATO planners.

Joint Exercise Decision Points

Some key decision points are also shown in Figure 2.7. The GAT met at 0800 hours and developed an apportionment recommendation, which if approved by the JFACC, was forwarded to the JFC. The JFC transmitted the AOOA to the JFACC at about 1300 hours. During the exercises the JTCB met only once at approximately 1400 to approve the JIPTL, which was then reviewed and approved by the JFC. At 1900 hours, shortly before the MAP was completed the JFACC approved the MAP, and at 2000 hours the JFACC published the Guidance and Intentions (G&I) message (also called the Air Tasking Message (ATM)).

At 0600 in the morning on the second day of the planning cycle the Naval Forces (NAVFOR) commander transmitted the Allocation Recommendation (ALLOREQ) message to the JFACC. The ALLOREQ provides the JFACC with a final estimate of the number of excess NAVFOR sorties available for the next day. Finally, the JFACC approved the ATO at 1300 on the second day of the planning cycle.

Chapter Three

AIR CAMPAIGN PLANNING IN PRACTICE

The discussion in the above chapter described how the air campaign planning process is supposed to work "on paper." In this chapter we examine how the process actually worked in ODS. A number of lessons-learned studies have been conducted which concern the air campaign planning process employed during ODS.[1] The following discussion draws on much of this previous work, including previous RAND studies. From this body of knowledge we have attempted to put together an integrated view of how air campaign planning worked in ODS.

From the observation of recent exercises we have also learned how the air campaign planning process was actually performed in these cases as well. However, it is difficult to separate the effects of training limitations and flaws in exercise design from actual limitations in the air campaign planning process. Therefore, we will not describe how the ATO production process actually worked in the exercises we observed.

ORGANIZATION AND ATO PROCESS

After the reorganization of the TACC, responsibility for air campaign planning was split between two organizations within the CPD: the GAT and ATO cells. The GAT was responsible for targeting and construction of the MAP and TPWs. Using the TPWs as inputs, the ATO cell provided the coordination and deconfliction necessary to turn the TPWs into a "flyable" ATO.

This organizational arrangement might have worked well if the ATO production process had proceeded in the carefully designed sequence of steps that constituted the process "on paper" and as described in the previous chapter. The actual process turned out to be significantly different and much more chaotic. Some of the disorder can be attributed to the chaotic nature of war and to the "fog of war" that can envelop wartime planners who lack necessary intelligence data or other needed information.

But a number of other factors contributed to the chaos that sometimes characterized the process. First, there were a relatively large number of changes made to the ATO at the last minute or when it was in the final stages of coordination and deconfliction. These changes made ATO production much more difficult, as inputs to the process

---

[1]A full list is not given here. The reader is referred to Cohen et al., 1993; and Joe and Gonzales, 1994.

were constantly changing. Second, ATO changes made by the GAT were sometimes not communicated to the ATO cell. The latter organizational problem made it even more difficult for the ATO cell to produce a "flyable" ATO.

> Changing the ATO while it was being coordinated, especially after it had been released to the wings, increased the complexity and fragility of the process.[2]

Because of these problems, getting the ATO transmitted to subordinate units on time was a major problem. In the first three weeks of the war, the ATO was published between 1800 and 2100.[3] The day 3 ATO was published before it could be completed. These delays reportedly prompted the JFACC to try to reduce the number of late ATO changes, but the wartime record of changes appears to indicate otherwise.

ATO changes resulted from several factors. The JFACC and General Glosson made every effort to use coalition air assets in the most responsive and flexible way possible. Prior to the reorganization of the TACC, ATO changes could not be made after 0800Z on the second day of the planning cycle. After the reorganization, the JFACC placed no absolute restrictions on when a change could be made to the ATO.[4]

After this change in policy the GAT made targeting changes at all stages in the process and sometimes within minutes of takeoff. The flow of target intelligence data into the TACC of course did not conveniently coincide with the ATO planning cycle. As new high-priority targets were identified, or as BDA became available on targets already struck, GAT targeteers tried to immediately update the ATO.

ATO CHANGES

With the responsive targeting approach adopted by the GAT, target or attack timing changes led to a considerable number of ATO changes. These types of ATO changes, the total sorties flown, and the total number of ATO changes for each day are shown in Table 3.1. Bad weather also resulted in a significant number of ATO changes (sortie cancellations), especially in the first week of the air campaign. The largest percentage of ATO changes occurred on day 5 when extremely bad weather was encountered and the TACC still suffered from tanker coordination problems. One thousand out of 2,300 sorties flown that day were ATO changes, but only 400 of the changes were weather related.

On average about 500 ATO changes were made per day. The average sortie rate for the entire war was 2,717 sorties per day. On average, about 20 percent of the sorties flown each day resulted from ATO changes, and about 8 percent resulted from target

---

[2]Cohen et al., p. 233.

[3]Gen. Glosson as quoted in Cohen et al., 1993, p. 233.

[4]Cohen et al., 1993, p. 208.

Table 3.1

Total Sorties Planned and Changed in Operation Desert Storm

ATO Day	Sorties Planned	Total Sorties Changed	Total Timing and Target Changes	Timing Changes	Target Changes
1	2759	0	0	0	0
2	2900	68	16	2	14
3	2441	449	112	36	76
4	2311	813	173	57	116
5	2286	975	207	83	124
6	2539	552	112	62	50
7	2803	687	211	43	168
8	2990	544	209	86	123
9	2657	531	121	41	80
10	2844	526	102	52	50
11	2555	604	171	67	104
12	3031	367	135	67	68
13	2914	220	81	70	11
14	2691	577	322	240	82
15	2859	543	281	139	142
16	2796	518	415	106	309
17	2607	488	214	51	163
18	2972	514	255	76	179
19	2856	650	273	108	165
20	3019	571	303	144	159
21	2581	612	251	169	82
22	2798	561	198	127	71
23	2929	433	293	102	191
24	2883	377	195	62	133
25	2854	426	99	34	65
26	2808	385	129	50	79
27	2863	363	195	23	172
28	2906	747	324	158	166
29	2778	488	240	79	161
30	2868	336	142	80	62
31	2656	530	242	39	203
32	2332	564	280	133	147
33	3158	369	204	42	162
34	3149	517	263	85	178
35	2580	629	326	132	194
36	2919	260	86	41	45
37	3119	667	454	107	347
38	3279	745	215	68	147
39	3309	718	350	73	277
40	3073	738	303	107	196
41	3271	905	307	89	218
42	2911	981	424	97	327
43	723	394	182	123	59
Total	116818	22942	9415	3550	5865

SOURCE: Cohen et al., 1993, Table 10, p. 243.

and timing changes. The other 12 percent of the sorties flown each day resulted from other ATO changes, including some weather-related changes. The cause of these other changes cannot be precisely determined. An unknown number of these were because of strike aircraft not finding needed tankers or escort aircraft at their rendezvous points—in other words because of breakdowns in the planning process.

However, on average about 40 percent of ATO changes, or 218 changes per day, were target and timing changes that were deliberately entered in the ATO. Most of these changes were made by the GAT, and they represent the realization of the JFACC's direction to use coalition air assets in the most responsive fashion possible. An example of this impetus concerns the redirection of F-117 strikes on a day when bad weather obscured the targets they had originally intended to attack. General Glosson reportedly ordered his deputy to find new targets (that were not obscured by the weather) for forty-four F-117s (nearly the entire deployed force) in 20 minutes, before their scheduled takeoff.

A tradeoff had to be made to achieve the level of targeting and attack flexibility observed during ODS. We have already identified how the ATO production process at the TACC was affected. Fighter-wing after-action reports indicate that late ATO changes also caused significant problems at the unit level. In some cases target changes occurred so rapidly, new targets were given to the wings without the requisite Desired Mean Points of Impact (DMPIs). Most wing mission planners were not trained to select DMPIs, nor were they fully aware of all the nuances involved in targeting or the subtleties involved with GAT objectives. This caused more time pressure during mission planning and led to unintended targeting errors.

In a number of additional cases, other equally vital targeting information was unavailable to the wings—target graphics or imagery. Target imagery was sometimes unavailable at the unit level even when no ATO changes were involved, primarily because of the very limited communications bandwidth available to the wings. Because of these communications limitations it was very difficult to deliver target materials for ATO changes to the wings on time, especially when new targets were selected within hours of takeoff. The F-15E lessons-learned report identifies late ATO changes as a significant detriment to mission effectiveness for these and other reasons.

> The time needed to plan AI missions is critical. Aircrews need to have ATO changes *at least 6 hours* before takeoff in order to plan properly. On several occasions ATO changes were received with little or no time to plan, brief, and upload appropriate munitions. Aircrews became less effective in executing interdiction missions. . . . Changes in the ATO should be the exception and changes . . . [that are] not time critical should be incorporated in the follow-on ATO.[5]

The F-117 community also expressed strong concerns about late ATO changes and suggested that they be abolished altogether.

_____

[5]USAF, 1991.

>   Amount of changes made in the ATO daily became almost overwhelming. . . . Rigid
>   rules need to be established at the TACC to prohibit last minute changes.[6]

The observations of these two wings do not appear to be unique. Similar observations were recorded in the F-111 wing lessons-learned report as well. It should be noted that a significant portion of the U.S. Precision Guided Munition (PGM) delivery capability employed in ODS resided in the F-117, F-111, and F-15E wings. The JFACC went to great lengths to employ these forces in the most flexible and effective way possible.[7] Consequently, these units may have been the focus of a much larger percentage of the ATO changes than other units. Earlier RAND research on this subject indicates this was indeed the case for the F-117 wing.[8]

TANKER PLANNING

The first two days of the air campaign were preplanned in detail outside of the normal ATO planning cycle. Detailed coordination for the preplanned ATOs was done relatively far in advance during Desert Shield. Consequently, the rendezvous of strike aircraft and tankers was calculated precisely and practically to the minute, and many contingencies were thought of ahead of time and compensated for.

The transition from preplanned ATOs to continuous ATO cycle operations was difficult for several reasons. Difficulties were especially encountered in tanker planning during the first week of the war. On day 3 many sorties had to be canceled because of tanker nonavailability.[9] On day 4 the TACC's ability to match tankers with strike aircraft declined even further. The dynamics of an air war involving hundreds of sorties an hour made it extremely difficult to optimize tanker employment with the limited tanker planning resources that were available.

Tanker planning was done manually in the TACC and was difficult to do with precision in the time available in the ATO cycle. It was even more difficult for personnel at the tanker operations desk in the TACC COD to compensate for late ATO changes during ATO execution. Figure 3.1, a highly simplified and unclassified depiction of the tanker tracks used, illustrates the scope of tanker operations during the Gulf War. In many cases tanker tracks had to be positioned in close proximity to one another—in both altitude and in latitude and longitude. Over 35 tanker tracks were used in the theater of air operations.

Because of the difficulties encountered with assigning tankers to specific strike aircraft, Gen. Glosson changed the tanker planning process after the first few days of the air campaign. Tanker planners no longer attempted to precisely match tankers to aircraft. Instead tankers were stationed on a continuous basis at tanker track positions. A portion of the tanker fleet was kept as an emergency reserve to fill unantici-

---

[6]USAF, 1991.

[7]Cohen et al., 1993, p. 228.

[8]Joe and Conzales, 1994.

[9]These cancellations are included in the timing changes and target changes categories change category in Table 3.1.

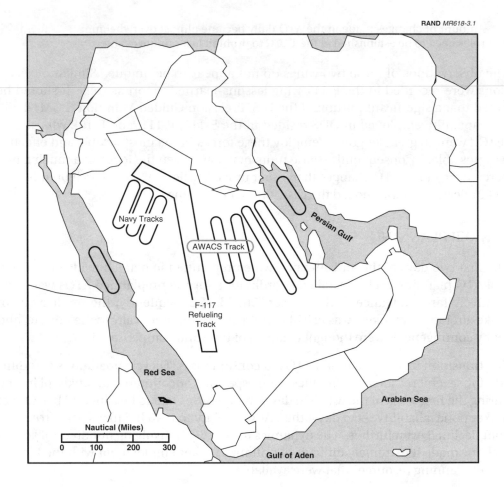

Figure 3.1—General Pattern of Desert Storm Tanker Tracks

pated shortfalls in tanker or fuel availability. This change simplified the overall ATO planning process. It reduced the load on tanker planners and made it easier to accommodate ATO target and attack timing changes. There was a danger however that tanker support could be unavailable at a specific location for a short period of time.[10]

It should be noted however that the simplified planning approach used for refueling operations may have made the associated real-time control (airspace management) task much more complex and difficult. Combat aircraft returned in groups and some aircraft ran dangerously low on fuel and needed immediate refueling. Because of the sheer size of combat air operations the airspace was at times extremely crowded in the vicinity of tanker tracks. AWACS controllers were pressed into an unanticipated airspace management role because of this problem and because of limitations asso-

---

[10]Because large amounts of jet fuel and ramp space were available throughout the theater of operations, little risk was incurred by adopting this concept of tanker operations. However in other potential theaters of war, this method of employing tanker resources may not be possible.

ciated with other parts of the TACS employed in the region. These problems illustrate the complexity of the ATO that was executed each day of the war.

ATO CHANGES AND ATO CYCLE TIME

Above we described how GAT personnel frequently had to circumvent the ATO production process because of the ATO cycle's 48 hour timeline. ATO changes were made which were disruptive to the overall planning process in the TACC and to mission planners at the unit level. However, the JFACC and the GAT accepted these disruptive side effects to carry out a responsive targeting strategy which maximized the utility of U.S. high-value attack assets.

The JFACC is presented with two conflicting goals: one, to produce a "flyable" ATO; and two, to maximize the responsiveness of available air power. The first goal is accomplished by deliberate planning to produce a fully coordinated and deconflicted ATO; the second, by maintaining the capability to make targeting changes at the last possible minute. How can these responsiveness and coordination goals be addressed simultaneously? With better automation support systems for ATO production, a well-trained AOC staff, and a clearly structured production process, it may be possible to reduce the length of the ATO planning cycle. Such a reduction in cycle time would be one particular way to address both issues simultaneously.

One key consideration should be kept in mind in reengineering the ATO production process. If the process is reengineered, the AOC organization must match the new process, and the individuals involved must understand how the process should work. We will not discuss AOC organization options in this report. We hope that the process improvement observations presented here can contribute to Air Force efforts aimed at improving AOC organization.

Analysis of the responsiveness issue suggests that responsiveness of the planning cycle can potentially be increased in four ways:

- Do away with the process and let the wings "do their own coordination."

- Shorten the cycle time associated with the process.

- Structure the process so a limited number of changes can be added at specific points in the process timeline.

- A combination of approaches two and three above.

Without going into the details of how such a scheme would work, we see that the first alternative, sometimes referred to as giving "mission type orders" to the unit level, is probably not feasible at this point—even if the latest commercially available collaborative planning environments were used. Such a scheme would require all participants to have near instantaneous access to a central planning database, and some form of supervision would still be required to deconflict individual unit-level plans. The scale of communications bandwidth required for such real-time collaborative planning environments will not be available in the near future to deployed military forces and very likely could not be acquired in the current constrained budget environment. This alternative also raises the issue of what type of awareness the JFACC should or would have of developing plans, and may require a complete rethinking of JFACC doctrine. We shall not consider this alternative further in this report.

The second alternative, shortening the cycle, will be explored in detail in the next chapter, but may not provide the increased responsiveness the JFACC may require against critical emergent targets. The third alternative is a way of permitting a limited number of ATO changes without introducing too much chaos into the deliberate planning process. The fourth alternative is a hybrid approach which preserves elements of the existing planning cycle but would allow new targets to replace old ones at certain points in a compressed planning process. We shall explore the last alternative as well below.

Compression of the ATO Planning Cycle

How much can the ATO planning cycle be shortened or compressed? To answer this question we examine the factors that may limit compression. The length of the timeline is determined by decisionmaking as well as planning activities. To compress the overall planning cycle, we must reduce the time needed to perform both types of activities.

First we consider the key decisionmaking activities. Recall that a number of key decisions are made by the JFACC and the JFC in the first 24 hours of the current cycle. The outcome of these decisions are necessary inputs to MAP and ATO production. These decisions or equivalent messages are the

- Initial JFC guidance message

- ALLOREQ message

- JFC apportionment decision (AOOA message)

- JTCB targeting decision (JIPTL)

- JFACC/GAT targeting decision

- JFACC G&I message.

For a compressed planning cycle, the majority of these decisions would have to be made near the beginning of the cycle. This is illustrated in Figure 3.2 with a notional ATO decisionmaking timeline for a 24 hour planning cycle. In recent joint exercises, the JFC AOOA message was not transmitted until about 7 hours into the cycle, which delayed the entire planning process significantly. During ODS the equivalent decision was made by CINCCENT about 3 hours into the cycle. If the cycle was compressed into a 24 hour period (a two-to-one compression factor), the JFC AOOA message should probably be transmitted about one and a half hours into the cycle (as illustrated in Figure 3.2). As indicated in the figure, a number of other decisions have to be made before the JFC apportionment decision can be made. The time needed to

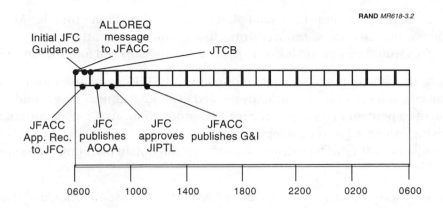

Figure 3.2—Notional Decisionmaking Timeline for a 24 Hour ATO Cycle

make these initial decisions would also have to be reduced by a factor of two to reduce the overall ATO cycle time from 48 to 24 hours.

Figure 3.3 indicates how the current ATO planning cycle could be compressed into a notional ATO cycle only 24 hours long. At least two additional modifications would be needed to enable such a compression of the planning process. First, the planning process must be further automated. Second, the overall planning process must be divided in an intelligent way into subprocesses which take less time to perform and coordinate.

Currently CTAPS automates ABP production and ATO compilation, and partially automates target development and weaponeering. Only MAP production remains a manual process. If an automated MAP production tool were developed and integrated with CTAPS, it could potentially speed up the process significantly. As illus-

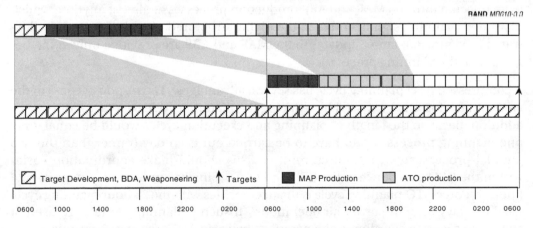

Figure 3.3—Compression to a Notional 24 Hour ATO Planning Cycle

trated in the figure, to cut the overall planning cycle in half, the time for MAP production would have to be reduced from the 11 hours currently needed to just 5 hours. This would be a reduction of 45 percent in the time used for MAP production.

Similarly, the time required for ABP production and ATO compilation would have to be reduced from the 13 hours currently needed to about 9 hours. This would be a reduction of 69 percent in the time used for ATO production. As CTAPS matures, as the underlying information technology advances, and as Air Force personnel become more proficient with this type of system, such a reduction in planning time should be achievable.[11]

Finally, if we assume that two hours are needed for ATO dissemination and that ATO production can be done in the time specified in Figure 3.2, planners at the unit level will have at least 6 hours available for mission planning. This would provide the minimum amount of time pilots say is necessary for mission planning.[12]

Now consider how subprocesses are organized in the overall planning cycle. Previous ATO cycle timeline charts depict a single serial set of processes. However, experience in ODS and in exercises has demonstrated that some subprocesses are carried out in parallel, while others must be carried in serial fashion during the ATO cycle. For example, before MAP production can start, a set of targets is needed. Similarly, before ABP production can begin, the MAP and target weaponeering options are needed. If targets are changed, the MAP and ABP must be modified. MAP, ABP, and ATO production must be carried out in serial fashion. However, target and weaponeering processes can be carried out in parallel with other planning activities. In fact, during ODS these processes were performed continuously throughout the cycle. These parallel and serial planning subprocesses are illustrated in Figure 3.3 by two separate tracks.

Target development and weaponeering must be linked to MAP and ATO production at least at one point in the cycle to provide a target list. During ODS, target changes were injected into the MAP and ABP production processes at almost anytime, which disrupted planning significantly. In the highly structured planning approach illustrated in Figure 3.3, target updates to the MAP and ABP are provided only at the beginning of the planning process.

A compressed ATO planning cycle has several advantages. The responsiveness of the *entire* attack force is increased by reducing the planning cycle by a factor of two. In addition, because the length of planning and execution cycles would be equal, only one planning process would have to be carried out each day (rather than the two parallel processes carried out currently). This should make coordination easier within the CPD and between the CPD and COD. Finally, another advantage of a 24 hour Air Force ATO planning cycle is that it coincides with the 24 hour planning cycle used by Navy aircraft carrier air operations. If such a planning cycle were used, it should be easier to coordinate air operations between Air Force and Navy units.

---

[11]This subject will be discussed in more detail in Chapter Five.
[12]USAF, 1991.

Compressed ATO Cycle with Prescheduled ATO Changes

Even with the compressed ATO cycle described above, a target detected early in the planning cycle may not be attacked for another 24 to 48 hours or until the ATO is executed. Ad hoc processes were invented during ODS to permit target changes to be made anytime within the ATO cycle, but these changes frequently were disruptive.

In the more structured target planning approach illustrated in Figure 3.4, ATO changes can be made at specific prescheduled points during the process. Beyond a certain point, when the ABP and MAP are in the final stages of coordination, potentially disruptive target changes would not be allowed.

After that point, disruptive target changes would be added to the next day's ATO instead. Such a planning process could allow the JFACC to pursue a more responsive attack strategy while preserving the benefits of a deliberate planning process.

The target changes indicated in Figure 3.4 are changes that could significantly impact coordination of the ATO. They would require additional coordination and planning before they could be included in the ATO. Such ATO changes could delay the ATO production process and make it more difficult to carry out a 24 hour ATO planning cycle if they occurred at random times in the process. It is envisioned that planners would be restricted to making such changes only at prescheduled points in the cycle, as indicated in the figure. During these times, work would stop on the central ATO databases, and ATO changes would be fed into the system. After all such ATO changes were loaded into CTAPS, CPD personnel would be alerted automatically by CTAPS to the impact of these changes on other parts of the ATO.

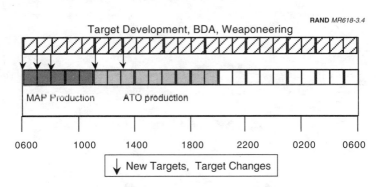

**Figure 3.4—Notional Schedule for Inserting ATO
Changes in a Compressed ATO Cycle**

Chapter Four

THE 5.0x CTAPS ARCHITECTURE

In this chapter we examine the current 5.0x CTAPS architecture. First we review the CTAPS hardware and software architectures, and then we examine how air campaign planning is currently implemented as a set of linear processes in CTAPS.

HARDWARE ARCHITECTURE

CTAPS runs on "open system" hardware. In CTAPS 5.0x, Sun Microsystems SPARC 2 and 10 workstations are configured as a client/server computer system. SPARC 10s act as servers and maintain most of the large databases used in the air campaign planning process. SPARC 2s act as clients on the CTAPS Ethernet local area network (LAN). SPARC 2s are relatively slow and have relatively little memory. CTAPS applications which use large databases require at least 16 MB and preferably 32 MB of RAM, and a 2 gigabyte hard drive to run efficiently.[1]

Because the CTAPS hardware architecture is based on state-of-the-art COTS products, it is dynamic in nature. The original CTAPS client/server hardware architecture was based on earlier versions of Sun Microsystems workstations that are no longer in production, and that soon may no longer be maintainable because replacement parts will be unavailable. As new more advanced open system COTS workstations are introduced they will have to be incorporated into the CTAPS hardware architecture, simply for maintainability reasons. This evolutionary acquisition approach to computer hardware is essentially the one adopted by commercial users, and it has become cost-effective as computer hardware has increasingly become a commodity.

Deployment Packages

The Air Force has developed three standard CTAPS deployment packages designed for various contingencies and AOC deployments: the quick reaction package, the limited contingency package, and the theater reaction package.

---

[1] This CTAPS hardware configuration was a typical one for mid-1994, when data were collected for this study.

The quick reaction package is composed of 29 CTAPS terminals and can connect up to 4 remote sites. This system will be able to operate for up to 14 days and can be used to plan ATOs containing up to 300 sorties a day.

The limited contingency package is composed of 99 CTAPS terminals and can connect up to 8 remote sites. This package will be able to operate for up to 30 days and can be used to develop ATOs containing up to 1,000 sorties a day.

The theater reaction package is composed of 144 CTAPS terminals and can connect up to 12 remote sites. This package will be capable of operating for more than 30 days. Using this system, AOC personnel will be able to plan up to 2,000 sorties per day.

All three packages are equipped with an Ethernet LAN sized to support the number of CTAPS terminals included in the package. These packages have been evaluated operationally in Blue Flag and overseas exercises, where CTAPS "stress tests" have become a regular occurrence. However, other, non-CTAPS, workstations could be deployed in future conflicts. It is not clear that the CTAPS LAN, the only LAN in the current AOC computer architecture, could support all workstations that could be deployed in future contingencies.

CTAPS is also now operational at various Air Force command and control centers around the world. The exact CTAPS configuration used is typically unique and typically varies from the deployment packages described above.

Modular Air Operations Center

CTAPS deployment packages are a part of a larger set of equipment which includes rapidly deployable shelters which expand to three times their stowed size. These shelters and the associated communications and prime power equipment form the Modular Air Operations Center (MAOC). The suite of MAOC communications gear includes Tactical Air Defense Data Links (TADIL) A and B. Eventually, a JTIDS link will be added as well. A CTAPS deployment package, MAOC, and associated communications gear will provide all the equipment necessary to establish an AOC in a contingency operation.

SOFTWARE ARCHITECTURE

The CTAPS software architecture is based on a combination of open system software, COTS-based software modules, government off-the-shelf (GOTS) software, and specialized mission applications. The three basic layers of the software architecture are depicted in Figure 4.1.

The bottom COTS-based software layer is based on a portable operating system (POSIX) compliant version of the Unix operating system (OS) and utilizes many open system software standards which have been adopted by DoD in the Defense Information System Agency (DISA) Technical Architecture for Information

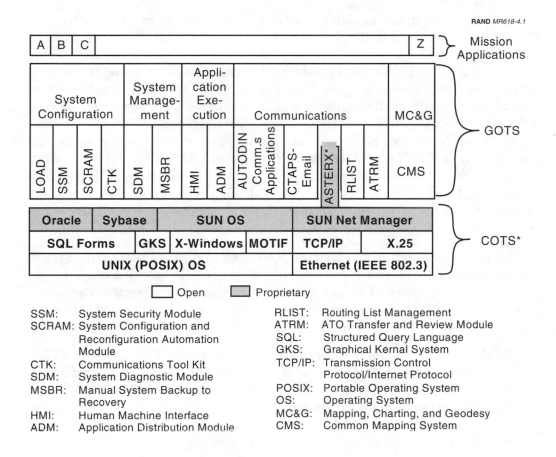

Figure 4.1—CTAPS Version 5.0x COTS/GOTS Architecture

Management (TAFIM).[2]   The TAFIM and open system software standards are discussed in more detail in Chapter Six. In addition to open system software, the bottom layer also contains the proprietary COTS software shown in the figure. The second software layer primarily contains GOTS software products for system management, system configuration, application execution, communications, and for the storage and display of mapping charting and geodesy data.

System configuration software is used to initialize the system.[3]

The CTAPS system management and applications execution modules are shown in the figure and provide self-explanatory services.

---

[2]The TAFIM compliant portions of the architecture are the POSIX compliant, Unix OS , the user interface standards X-windows and Motif, the Graphical Kernal System, Ethernet, and X.25.

[3]LOAD loads the specific repertoire of software modules needed on each workstation. The Communications Tool Kit tailors network interfaces for printers and workstations. SCRAM positions and initializes databases on workstations and hard drives. It can also recover databases if such a need should arise. The SSM identifies security labels and provides duty group authorizations and system audit parameters.

CTAPS communications modules provide a range of services. CTAPS has two Electronic Mail (E-mail) facilities, a unique GOTS E-mail application, called CTAPS E-mail, and a COTS E-mail product called ASTERX which is built into a CTAPS mission application. CTAPS E-mail provides some capabilities not found in standard COTS E-mail products, such as the ability to send E-mail messages to duty stations, instead of individuals (individual login accounts). But problems have been encountered with CTAPS E-mail in some recent exercises. It has proven difficult to configure the application correctly and E-mail messages have proliferated and have made data transmission on the CTAPS LAN difficult. In addition, CTAPS E-mail is not interoperable with other COTS E-mail products that use standard open system E-mail protocols. In contrast, ASTERX is compatible with open system E-mail protocols and has proven useful in providing E-mail communications to non-CTAPS workstations.[4]

CTAPS provides other communications services besides E-mail. The System Message Alert (SMA) module sends small messages to CTAPS workstations (e.g., system shut down warnings). The Routing List Management (RLIST) program generates duty station position lists, and the ATO Transfer and Review Module (ATRM) transfers ATOs to remote CTAPS terminals.

CTAPS also uses Automatic Data Interchange Network (AUTODIN). The CTAPS AUTODIN architecture is illustrated in Figure 4.2. AUTODIN messages conforming to the U.S. Message Text Format (USMTF) can be constructed from CTAPS data, routed, and transmitted using the applications shown. AUTODIN messages can also be received and loaded into CTAPS databases. AUTODIN messages can be transmitted to AUTODIN Switching Centers as shown in the figure.

AUTODIN communications are used for record message traffic, such as JFC guidance messages or Operations Orders. During ODS and in recent exercises, AUTODIN was used to transmit the ATO when no other means was available or when problems were encountered with CTAPS network communications with remote terminals. In many cases, ATO transmission via AUTODIN led to significant problems within the AUTODIN communications network. Large message backlogs were encountered at AUTODIN Switching Centers, and AUTODIN message centers on Navy ships were caused to crash. AUTODIN was not designed for the transmission of large or even moderately sized ATOs.

Another important part of the CTAPS GOTS application layer is the Common Mapping System (CMS). It provides mapping services to a number of CTAPS mission application modules.

---

[4]The capabilities of CTAPS E-mail could be mimicked by using the alias feature of a standard COTS E-mail system. Individuals would have to be added and removed from the alias as they went on and off station in order for them not to be deluged with E-mail when they came on station. This would introduce extra work for system administrators but would allow a well-tested and interoperable COTS product to be used. There may also be other COTS E-mail products available that can provide the needed capability in a more efficient and transparent fashion.

RAND *MR618-4.2*

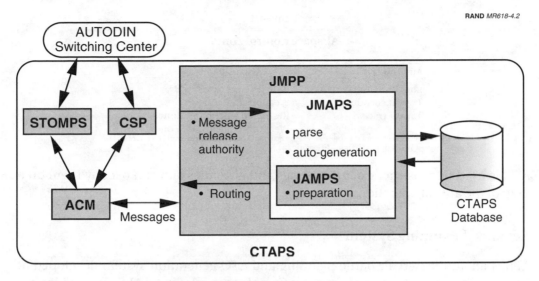

ACM: AUTODIN Communication Module
CSP: Communications Support Processor
JMAPS: Joint Message Analysis and Processing System
JMPP: JINTACCS Message Processing and Preparation
JAMPS: JINTACCS Automated Message Preparation System
STOMPS: Stand-alone Operational Message Processing System
Nb: CSP used only in HTACC in Korea. Other CTAPS installations use STOMPS.

Figure 4.2—CTAPS 5.0x AUTODIN Architecture

CTAPS VERSION 5.0x APPLICATION MODULES

The top software layer of the CTAPS 5.0x software architecture contains the mission application modules shown in Figure 4.3. These CTAPS mission application modules are described below.

Airspace Deconfliction System (ADS)

ADS is used to construct the Airspace Coordination Order (ACO) and utilizes CMS to provide a graphical representation of the ACO. The ACO divides the airspace in a combat zone into the areas listed in Table 4.1. ADS also displays tanker tracks (air-to-air refueling points) and package coordination or rendezvous points.

RAND *MR618-4.3*

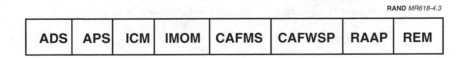

Figure 4.3—CTAPS Version 5.0x Mission Application Modules

Table 4.1

Airspace Control Zones

Air Routes	Exclusion Zones
Low-level transit routes	Base defense zones
Transit corridors	Weapons free zones
Tanker tracks	Restricted Operations Zones
	Weapons Engagement Zones

SOURCE: Griffiss AFB, n.d. (a).

ADS is used to construct exclusion zones and air routes and can overlay them on a map of the area of operations.

Advanced Planning System

APS is an automated air battle planning and ATO generation system developed at Rome Laboratory. It is one of the more complex modules in CTAPS. It provides automated planning support for strike (or target), tanker, reconnaissance, escort, ground alert, and orbiter missions.[5] It can assist the planner by performing mission feasibility, route, and mission flow analyses. It also checks for semantic and syntactic errors when data are entered into the APS database.

APS primary capabilities are in four functional areas: ATO management, database management, data import, and air battle planning. We briefly review these capabilities.

ATO Management. APS is used to manage a full set of ATOs. It can create, modify, and delete ATO databases. It can also maintain an archive of past ATOs.

Database Management. APS uses a number of databases. An APS ABP shell consists of three layers of data: theater data, scenario data, and ABP data. Theater data are seldom changed and must be entered into the system before ATOs can be prepared. APS theater data types are shown in Table 4.2. For example, if an adversary is known to have MIG-29s in its inventory, this aircraft type would be included in the APS theater database.

Table 4.2

APS Theater Data Types

Aircraft Types	Missile Equipment
Mission Types	Jammer Equipment
Standard Conventional Loads	Radio Equipment
Radar Equipment Air Bases	Digital Map Data

SOURCE: Griffiss AFB, n.d. (a).

---

[5]Orbiter missions are those in which the mission aircraft is assigned to a specific orbit location. Examples are defensive counter air missions assigned to a combat air patrol orbit, or AWACS aircraft assigned to a specific surveillance orbit.

Scenario data typically change daily and are used to prepare APBs. This type of data is illustrated in Table 4.3. Many of the data types are self explanatory. Logistics data refer to aircraft and munitions availability data for specific air bases. Airspace data refer to the ACO, and guidance data refer to the AOOA and other guidance messages.

ABP data are specific to a single ABP or ATO, such as package and target assignments for specific aircraft. ABP data change daily but may change more frequently as the planning process progresses.

Data Import. APS can import scenario data from other CTAPS mission applications. The data APS can import and the source mission applications are shown in Table 4.4.

Air Battle Planning. APS models six types of aircraft missions: target, reconnaissance, tanker, orbiter, escort, or ground alert missions. On-screen worksheets to plan each type of mission are provided. The details of strike package coordination are arranged using these worksheets, where tankers, call signs, and IFF codes are assigned to aircraft.

APS has an ABP deconfliction tool which automatically checks the timing assumptions entered for each air mission in the ATO. Its calculations are based on aircraft airspeeds, routes, TOTs, and other data. If the deconfliction tool arrives at a contradiction in mission timing, it issues a warning message regarding mission feasibility.

Other APS automation aids are the Autoplanner, a route planning tool, and an electronic combat (EC) analysis tool. Specialized CTAPS mission applications are often used instead of these APS tools if high or moderate fidelity results are needed. The APS route planning tool is especially useful for tanker planning.

Table 4.3

APS Scenario Data Types

Logistics	Weather
Intelligence	Guidance
Targets	Tactical Data
Airspace	

SOURCE: Griffiss AFB, n.d. (a).

Table 4.4

APS Data Import Capabilities

Data	Source
Enemy OB	ICM
Equipment	ICM
Coordination/ Rendezvous Points	ADS
Airspace Control Zones	ADS
Target Nomination List/ Weaponeering Options	RAAP

SOURCE: Griffiss AFB, n.d. (a).

APS EC and route planning tools use the CTAPS CMS. The graphics and computer intensive nature of these tools can significantly reduce the execution speed of APS—especially with large ATOs, so planners have tended to avoid their use during exercises. When more powerful computer hardware becomes available, the graphics-intensive capabilities of APS may be used more often.

The APS Autoplanner can assign tankers to aircraft, assign aircraft to targets, and perform nearly all the calculations necessary to complete an ABP. The Autoplanner does have limitations however. Its calculations are based on simple routes, which include only the minimum number of way points dictated by the ACO, package rendezvous, and air refueling points. Missions are assigned to targets by using a set of control parameters and priority weights which must be entered beforehand. The Autoplanner is optimized to finish planning already started. It is not designed to be used at the start of the planning process or when specific TOTs are required.

One important advantage APS provides is a relatively user-friendly graphical interface. For example, in constructing a strike package, the planner can open a window displaying all available strike aircraft. He can open a second window containing the strike missions already in the ATO. By simply "pointing and clicking," he can create new strike missions by removing entries from the first window and adding them to the second.

Computer Assisted Force Management System

CAFMS was originally developed as a stand-alone system with its own hardware and was designed to provide automation support for the CPD and COD of an AOC. It was used during the Gulf War for ATO production and dissemination. At that time it was hosted on obsolete computer hardware. Numerous difficulties were encountered with the system during ATO production and dissemination during Desert Storm.

CAFMS is now hosted as a separate mission application module in CTAPS 5.0x. It is an interim capability that will be phased out when version 6.0 becomes operational. It is currently used in the CPD as a communications program for ATO dissemination, and in the COD as an automation aid.

The CPD division uses CAFMS to collate the ATO (combine the ACO, ABP, and SPINS), to put it into the correct USMTF format, check it for formatting errors, and to transfer it to the CTAPS JINTACCS Message Processing Program which then transmits it into the AUTODIN network.

The CPD uses CAFMS as a database management tool. CAFMS can be used to query the published ATO database (the ATO being executed) to determine which aircraft are on ground alert, which targets will be attacked in the next hour, and which aircraft are in the air and can be diverted, etc. These capabilities are useful for real-time battle management and deconfliction.

CAFMS is also used by COD personnel to maintain and update databases on aircraft, munitions, airfields, air defense weapons, communications circuits, and air crew status. Returning air crews enter Mission Reports (MISREPs) into CAFMS. MISREPS are transmitted back to the AOC and are used to update CTAPS intelligence, targeting, and logistics databases.

Combat Air Force Weather Support Program (CAFWSP)

The CAFWSP can import and display a variety of weather data. It can display current and forecast weather maps, areas of cloud cover, Visual Flight Rules (VFR) areas, and Instrument Flight Rules (IFR) areas. It also can store and display visibility, wind, and precipitation data, and air base weather observations and forecasts. CAFWSP displays and imports or exports weather maps using the Army Uniform Guided Data Field (UGDF) data format.

Intelligence Correlation Module

ICM is designed to support the intelligence analyst in the correlation of intelligence data and the production of OB databases. It is preloaded with parts of the standard extended intelligence database (XIDB) information before deployment from garrison.[6] ICM is used to maintain friendly and enemy OBs, including aircraft, ground force, facility, installation, and electronic OBs. ICM uses the CTAPS CMS to display OB data.

ICM is equipped with a windows-based interface. Intelligence personnel can use it to quickly search OB databases according to location, equipment, type of facility, military units, etc. In the current version of ICM, version 1.0, OB databases must be updated manually, and it cannot be used to receive or process imagery. Version 1.0 also does not have an automated interface with Constant Source (CS) or any other near real-time intelligence dissemination system. At present the only communications interfaces available to version 1.0 of ICM are those listed in Figure 4.1 of the CTAPS software architecture. It can use the suite of CTAPS AUTODIN communications modules to transmit intelligence reports to external agencies and units, and it can receive intelligence reports from external sources via AUTODIN.[7] Future versions of ICM will address these shortcomings and will be described later in this report.

ICM can receive and display electronic OB (EOB) data generated by Improved Many on Many (IMOM), and it can automatically transmit OB databases to RAAP. ICM can also interface directly and share Intelligence Database (IDBs) with Sentinel Byte, the unit-level intelligence system that is part of the Wing Command and Control System (WCCS).

---

[6]The database used is the Defense Intelligence Agency (DIA) Military Integrated Intelligence Data System (MIIDS).

[7]There are upgrade plans to improve ICMs' external data exchange capabilities; however at the time this report was being written these planned improvements were not funded.

Improved Many on Many

As its name implies IMOM is an electronic combat assessment tool that can incorporate the effects of multiple jammers, radars, and aircraft. It can perform relatively high-fidelity simulations of the EC environment in the presence of multiple threats. It is used to assist in route planning, strike package planning, and EC planning.

Rapid Application of Air Power

The current version of RAAP provides automation support for target development and weaponeering. It can run as a stand-alone application that can operate in a system high environment or as a CTAPS mission application. When used in the CTAPS environment, RAAP operates only at the secret level. RAAP currently provides automation support for the following targeting functions:

- Target identification and characterization

- Vulnerability analysis and aim point selection

- Weaponeering

- Target nomination, and

- Bomb damage assessment.

In addition, RAAP currently provides limited support for BDA of strategic targets.

Target Identification and Characterization. RAAP currently has the capability to accept a variety of targeting information, including text-based target reports, Intelligence Summaries (INTSUMS), and a variety of imaged-based targeting products.[8] RAAP can incorporate two dimensional digitized drawings and soft copy imagery from a number of sources (e.g., national imagery from the 480th Air Force Tactical Intelligence Group, and LANDSAT or SPOT imagery). In addition, it can associate imagery products with targets listed in various IDBs. Using these capabilities, intelligence analysts can identify and characterize targets.

RAAP has been designated to maintain the CTAPS master target database, including the status, position, cover, definitions, and relative priority of all targets of strategic importance. RAAP will not physically maintain its own target databases separate from the OB databases maintained by ICM. There are plans to link RAAP targeting and ICM OB databases by means of associations that RAAP establishes between target imagery, other material, and ICM OB data objects.

A full range of intelligence OB data can be imported from ICM, although only complete, entire OB databases can be imported with the current versions of these application modules. OB data updates cannot be automatically sent from ICM to RAAP (i.e., the databases are not automatically linked).

---

[8]RAAP can accept and display soft copy versions of Annotated Targeting Graphics (ATTGs), Basic Target Graphics (BTGs), images in National Imagery Transmission Format (NITF), and images in other formats.

Vulnerability Analysis and Aim Point Selection. With the imagery import and analysis capabilities of RAAP, this application module can also be used to assess the vulnerabilities of various types of strategic targets. RAAP also can be used to select DMPIs.

Weaponeering. Intelligence analysts can weaponeer targets in RAAP by using an automated on-line version of the Joint Munitions Effectiveness Manual (JMEM). Currently, only the effectiveness of single weapon attacks can be modeled. RAAP provides on-screen target planning worksheets similar to those used during ODS to support the weaponeering process.

Some users have suggested that RAAP's weaponeering capabilities be expanded to increase the planning flexibility available to targeteers. In the current version of RAAP a total of only five DMPIs can be chosen per target, and only three weaponeering options can be attached to each target. These limitations are not an issue when weaponeering simple targets, but they can make the targeting process difficult in the case of complex targets. If these limitations can be removed they will also give ATO planners more attack aircraft options to choose from and will increase the flexibility of the ATO production process.

Target Nomination. One of RAAP's primary functions in the air campaign planning process is production of a fully weaponeered Target Nomination List. The TNL serves as the basis for ATO production, and it must be transferred to APS before detailed ATO production can begin.

Bomb Damage Assessment. RAAP can also provide limited support to the Bomb Damage Assessment (BDA) process. The operator can add BDA entries to targets in the master target database and maintain a history of the target.

Route Evaluation Model (REM)

REM is a specialized CTAPS application used for route planning. It can automatically accept IMOM data and can be used to interactively plan ingress and egress routes for threat avoidance. The results of REM runs are not passed to mission planners or pilots at the unit level. REM results can be used by force-level planners during the planning process. More precise route planning is done at the unit level using other route planning systems that are tailored to the capabilities of particular aircraft.

Chapter Five

THE AIR CAMPAIGN PLANNING TOOL AND RELATED INTELLIGENCE SYSTEMS

A number of other automation aids have been developed to support air campaign planning and intelligence analysis at the force level. Some provide automated planning and intelligence support capabilities that CTAPS currently cannot provide. We briefly review the capabilities of these potentially complementary systems below.

AIR CAMPAIGN PLANNING TOOL

ACPT is a force-level air campaign planning tool developed under an Advanced Research Projects Agency (ARPA) contract for the Air Force. It is a JFACC decision aid which can support the rapid development and amendment of air strategy options and be used to generate multiple-day strategic air attack options.

One of the central features of ACPT is its target database which contains correlated intelligence data for a complete set of targets for a wide variety of potential adversaries. The security level of the entire ACPT system is determined by the classification of the target database. It can operate at secret and higher classification levels.

ACPT employs a strategy-to-tasks (STT) approach as illustrated in Figure 5.1. In the STT framework, U.S. national security goals are used to derive military, foreign policy, economic, and political objectives for a particular scenario. These high-level objectives are then linked to the JFC's military campaign objectives for a specific theater of war and scenario. These campaign objectives are in turn used to derive a set of JFACC's air campaign objectives, and finally an air campaign plan.

The STT framework provides a link or "audit trail" connecting the JFACC's air campaign plan with the strategic objectives of the National Command Authority (NCA) and JFC. Using these links the ACPT assists planners by constructing a prioritized target list that conforms to the operational objectives needed to fulfill the JFACC's air campaign plan. The priorities assigned to targets ensure that NCA and JFC military objectives are achieved in the desired sequence of steps.

ACPT can be linked to a full set of classified national intelligence databases. Such linkages permit the ACPT prioritized target list to be based on the latest most accurate data but present a system integration challenge because CTAPS is designed to operate at the secret level.

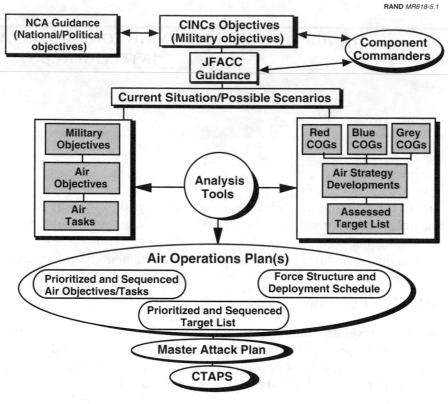

RAND *MR618-5.1*

COGs: Centers of gravity.

SOURCE: USAF, n.d.

Figure 5.1—Air Campaign Planning Tool Strategy-to-Tasks Approach

ACPT software runs on open system hardware (Sun SPARC workstations). The current ACPT hardware suite is illustrated in Figure 5.2. Although ACPT applications and databases are currently not compatible with CTAPS (the two systems cannot be linked together electronically), a floppy disk interface has been developed for transferring the Master Attack Plan from ACPT to CTAPS. The ACPT MAP production capability is based on an optimal weapons allocation simulation tool called the Conventional Targeting and Effectiveness Model (CTEM).

The prioritized target list produced by ACPT is input into CTEM, along with munitions and aircraft availability data. Weapons are allocated to targets using a linear optimization process with constraints. The mathematical constraints are in turn derived from the high-level attack goals specified in the STT set of objectives defined earlier in the ACPT planning process.

The CTEM module in ACPT can derive a notional sequence of MAPs for a five day air war (which are about 80 percent optimal) for a given strategy and set of attack resources within about half an hour. It is estimated by CHECKMATE personnel that this relatively short response time would allow planners to produce an almost opti-

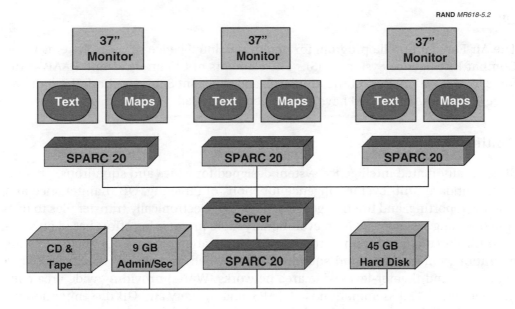

SOURCE: USAF, n.d.

Figure 5.2—Current ACPT Hardware Suite

mal MAP within five hours.[1] This would cut the time required for MAP production in half and would provide a significant enhancement to the JFACC's air campaign planning capabilities.

One limitation of the current ACPT MAP production capability (CTEM) is that it does not optimize the use of air refueling assets or reflect their availability in assigning aircraft to targets and aircraft to packages. However, in the current manual MAP production process, air refueling constraints are taken into account only very approximately by planning personnel. During Desert Storm, both MAP and ABP planning were hampered because the use of air refueling assets could not be optimized (no reliable automated decision aids were available). Today the APS route planning in CTAPS does provide air refueling automation support during the ABP process.

RELATED INTELLIGENCE SUPPORT SYSTEMS

A number of intelligence support tools have been developed by the Air Force and joint commands (or agencies) that are planned for use in the AOC, or that will connect remotely to intelligence support systems in the AOC. Some of these are integrated into the CTAPS architecture, while others are not.

---

[1]The current ACPT MAP production process does not take into account aerial refueling needs, so in this sense the MAP is not optimal. However, it should be noted that in the current manual MAP production process, aerial refueling needs are dealt with only approximately.

CIS

The Air Force umbrella program for deployable intelligence support systems is the Combat Intelligence System (CIS). The elements of CIS are ICM and RAAP at the force level, Sentinel Byte at the unit level, and Constant Source at both the unit and force levels. RAAP and ICM have already been discussed.

Sentinel Byte (SB)

SB is an automated intelligence system designed for wings and squadrons. It supports standard unit-level intelligence functions of targeting, OB maintenance and display, reporting, and briefing generation. It can electronically transfer files to mission planning systems. SB is an evolutionary rapid prototyping effort based on standard Air Force open system hardware. SB workstations will be connected to wing operation center (WOC) and squadron LANs which in turn will be connected to wing-level and theater-level wide area networks (WANs) providing wide area connectivity for ATO dissemination via CTAPS, and imagery and OB dissemination via digital ICM-SB links.[2]

When this document was in preparation the CIS architecture had not yet been completed. Recent related joint developments and difficulties in funding the acquisition of needed LAN and WAN connections have complicated and delayed completion of the architecture.

Constant Source

CS is composed of the two subsystems shown Figure 5.3: a receiver suite and an operator terminal. The operator terminal performs message filtering, track correlation, and display functions.

When CS is connected to a UHF satellite communications terminal it can receive and process Tactical Receive Equipment and Related Applications (TRAP) and Tactical Information Broadcast Service (TIBS) messages. Currently, the CS operator terminal is based on standard Air Force open system hardware. It is equipped with automatic message processing and correlation capabilities, so new contact reports can be automatically added to the CS OB database. In addition, the operator can set the CS message processing filters so that the system will automatically incorporate and display information on only the type of targets or threats specified by the operator.

At the present time, CS is a stand-alone system and is not integrated with CTAPS. There are plans to integrate it with CIS and to add LAN connectivity to CS.

---

[2]Prowse, n.d.

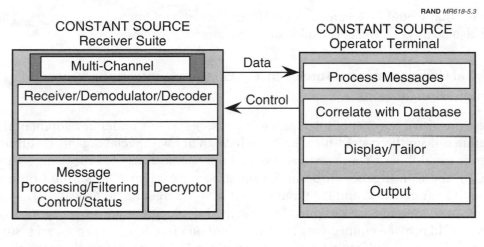

RAND *MR618-5.3*

SOURCE: USAF, 1990.

Figure 5.3—Constant Source System

Joint Intelligence Support System

JDISS is a stand-alone workstation client tool developed by the DIA and DISA. It is based on open system hardware and can access the MIIDS XIDB located at joint and theater intelligence centers.[3] Two versions have been developed for secret and system high-level access. JDISS is designed so client workstations can access IDB databases servers remotely over a WAN. It is equipped with the Secondary Imagery Graphics System (SIGS) and can download digital imagery files.

Some remote access problems have been encountered in recent exercises because of limitations in the JDISS client/server network design. Even though JDISS is based on open system hardware and software, its user interface demands complex command line prompts and responses. It does not have a windows-based user interface and has introduced training problems. Air Force intelligence analysts have tried to use the JDISS SIGS application for targeting; however the quality of SIGS imagery was deemed not to be high enough for this application. There are plans to remedy these difficulties.

5D Workstation

This imagery workstation was developed by the Central Imagery Office (CIO) and DISA. Currently the 5D client tool is a stand-alone workstation system based on open system hardware and can operate only at the system-high level. The current 5D client system is a part of another stand-alone WAN-based client/server system. 5D client machines can download secondary and primary imagery from CIO imagery

---

[3]The MIIDS XIDB is accessible remotely by on-line communications over the Defense Information Systems Network (DISN).

servers that would be located in CONUS and at regional intelligence centers. They can also be used for imagery exploitation and targeting.

Special Compartmented Intelligence (SCI) Communications and Exploitation Systems

There are also a number of intelligence processing systems under development that operate only at the system-high level and have their own special organic communications capabilities. These systems are the Joint Service Imagery Processing System (JSIPS), the Contingency Airborne Reconnaissance System (CARS), the Common Ground Station (CGS), and the Combat Intelligence Targeting on Arrival (CITA) van. All of these systems are designed for the rapid deployment contingency mission. Each would provide connectivity to a number of different intelligence sources; some use special communications links; and some provide direct connectivity to special assessment centers in CONUS.

A common difficulty encountered with each of these systems is that none are equipped with MLS interfaces. Consequently, none of these systems can be directly connected with CTAPS or other secret-level C4I systems. This has made it difficult to include them in larger Theater Battle Management (TBM) architecture planning efforts.

Chapter Six

THE PLANNED CTAPS 6.0 ARCHITECTURE

The next planned version of CTAPS software, version 6.0, will incorporate a number of new mission application modules, improvements to existing modules, and near-real time information processing and display capabilities. The suite of hardware equipment for CTAPS 6.0 is essentially the same as that used in the current architecture, except for ICM.[1]

The new mission applications modules that are planned for CTAPS 6.0 are the Force Level Execution system, the BSD, and ACPT. One existing mission application module, CAFMS, will be removed. In the COD, CAFMS will be replaced by FLEX, and in the CPD, it will be replaced by APS.[2] APS will be modified to perform the ATO collation and dissemination functions now performed by CAFMS, and the capabilities of the RAAP module may be expanded as well.

POTENTIAL RAAP IMPROVEMENTS

RAAP is an ambitious evolutionary acquisition program that may some day be capable of providing a wide range of automated support functions to intelligence analysts and targeteers. Many of RAAP's more powerful potential capabilities have yet to be implemented and would be based on AI technology. In addition, there are plans to upgrade the weaponeering support capabilities of RAAP.

Weaponeering

In the current version of RAAP, only the effectiveness of single weapon attacks can be modeled. However, when development of the Advanced Weaponeering Optimization Program (AWOP) is finished, it will be integrated into RAAP. RAAP will then have the capability to estimate the effectiveness of multi-aircraft and multi-weapon attacks against targets.

Users have suggested that RAAP's weaponeering capabilities be expanded to increase its flexibility. In the current version of RAAP a total of only five DMPIs can be chosen

---

[1]The current version of ICM, version 1.0, only runs on SPARC 2 workstations. Version 2.0 of ICM, which will be incorporated into version 6.0 of CTAPS , will run on the SPARC 10.

[2]CAFMS will still be used at the unit level. It will be ported to the WCCS and used by mission planners to receive and process the ATO.

per target, and only three weaponeering options can be attached to each target. These limitations are not an issue when weaponeering simple targets, but they can make the targeting process difficult for complex targets. If these limitations can be removed they will also give ATO planners more attack aircraft options to choose from as well and will increase the flexibility of the ATO production process.

Potential Future Capabilities

A number of desirable RAAP functions have been deferred because of funding constraints. These potential capabilities include automatic target database maintenance. As new Target Bulletins (TARBULs) are received, they would automatically be processed by RAAP, and the new targets identified in TARBULs added to the RAAP target database. The system would also automatically enter scheduled target attack data from the ATO, including expected levels of target damage. RAAP would subsequently access the released ATO and update the target database as scheduled target attacks are carried out or deferred.

Currently, RAAP has limited automatic message processing capabilities. It can accept a wide variety of imagery data, AUTODIN message traffic, other text-based messages, and ICM OB databases. However, only the latter data can be automatically processed and stored in RAAP databases. Specific data elements from text based messages must be manually identified and entered in RAAP databases. Planned enhancements would add automatic message processing features to RAAP.

Another potential RAAP capability is an automatic electronic interface with external collection management and collection requirements systems, such as the Swift Hawk II Collection Requirements Management System.

A significant amount of research has been carried out by RAAP contractors and others using AI techniques to automate decisionmaking and inference processes related to the interpretation of intelligence information. A set of possible future RAAP AI capabilities would integrate high-level knowledge of enemy operations with current and historical data to provide estimates and predictions of enemy military activity. These predictions would be used to identify high-value targets and to recommend effective ways of using air power assets.

FORCE LEVEL EXECUTION SYSTEM

In the CTAPS 5.0x architecture CAFMS provides very limited automation support to COD personnel—only very basic database storage and retrieval capabilities. FLEX will provide more advanced database management capabilities and sophisticated decision support tools for COD personnel. FLEX is the product of two separate programs: a 6.3A Advanced Technology Transition Demonstration (ATTD) program, and a 6.3B effort that will deliver rapid prototype systems to the field. Both are being developed at the Advanced Concepts Branch of Rome Laboratory with significant operator input from various Numbered Air Forces, Pacific Air Force, and U.S. Air Force in Europe.[3]

FLEX is designed to fit into version 6.0 of CTAPS as a mission application module. It will interoperate with other CTAPS mission application modules and with selected external systems. It can be configured to allow up to 30 users to work in parallel on separate workstations, using the same ATO databases.

FLEX will also reportedly be capable of supporting COD operations in a compressed ATO cycle.

> It will also allow combat operators to effectively manage force level air operations of over 3000 sorties during a single ATO execution period, even if the current ATO execution cycle is decreased in the near future.[4]

In particular FLEX will provide advanced automation support for COD personnel to

- coordinate, integrate, and control current theater air operations

- understand air mission relationships established in the ATO

- maintain current force status, enemy OB, mission, target, and weather databases

- monitor the activities of subordinate TACS elements

- adjust air taskings in response to changing battlefield dynamics

- prepare and disseminate Change Task Orders (CTOs)

- summarize and report mission execution results.

FLEX will support these functions by providing the ability to: one, monitor ATO execution; two, alert duty officers and combat operators of potential problems during execution; and three, automatically generate replanning options to fix problems.

FLEX ATO monitoring and deviation detection tools will automatically access the FLEX published ATO database. The published ATO database will be kept updated by a status reporting facility built into FLEX. Status reports will be received from the units either automatically (MISREPS) or verbally. Verbal status reports will be input manually into FLEX. As a mission is executed a series of status reports will be entered into the FLEX published database to keep it current with executed operations.

At the present time there are no plans to incorporate real-time air picture data into FLEX databases.

ATO Monitoring Tools

In initial versions of FLEX, two types of ATO monitoring displays will be provided, a status display board, and the Marquee. Later versions of FLEX may include an ATO

---

[3]Clark, 7 June 1994.
[4]Clark, 7 June 1994.

"animation" capability in which the position and speed of air missions will be simulated using ATO data. Animated air missions would take off from air bases, rendezvous with other package members, fly specific waypoints, and attack targets according to the TOTs listed in the ATO. The FLEX ATO simulation capability may require significant computer resources to run in the faster-than-real-time modes that will be needed for it to be of use to COD personnel. Thus, it may only be useful when it can run on high-speed next generation workstations.

The FLEX Status Display Board is a tabular representation of the manual "greaseboards" used in the AOC today. It will provide a list of air missions, including ATO information such as TOTs, tanker rendezvous times, mission type, on-station times, etc. This display can be tailored according to duty position, so for example, the tanker duty officer can display only tanker missions. The Status Display Board is tied directly to FLEX databases and does not have to be updated manually.

The Marquee provides a graphical representation and lists air missions in the published ATO database. It is linked to the published ATO database and allows users to monitor ATO execution. The Marquee can display temporal relationships of missions in a Gantt chart format, as shown in Figure 6.1. It is automatically updated as changes are made to the ATO.

The Marquee can be used to query the database using a number of search strategies and an intuitive user interface (without having to use complex SQL commands). A mission in a strike package can be extracted from the database, including all neces-

RAND *MR618-6.1*

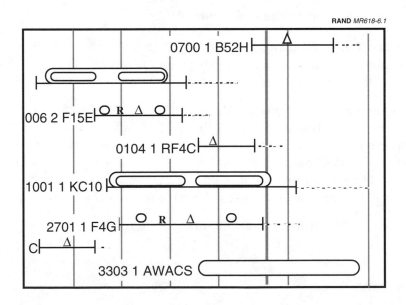

SOURCE: LaBatt, n.d.

Figure 6.1—Marquee Graphical Display

sary refueling and escort aircraft.[5] If a user requests all aircraft in a strike package, he can bring up this data by selecting a single mission known to belong to the package. FLEX will automatically generate the multiple database queries to construct the required list of missions without further user intervention.[6]

The meaning of Marquee icons are indicated in Table 6.1. Key mission milestones and estimated flight times are represented in the Marquee display. It will permit COD personnel to quickly ascertain detailed relationships between mission aircraft. Currently with CAFMS, operators must search through rows of text-based mission data to ascertain such relationships. The FLEX graphical user interface (GUI) will greatly simplify database search tasks.

Alerting and ATO Deviation Detection Tools

FLEX will be able to automatically generate alert messages to notify combat operators of potential conflicts that may arise during ATO execution. As mentioned above, as a mission is executed status reports are entered into the FLEX published ATO database. FLEX will automatically check mission update messages to determine whether they impact other missions scheduled in the ATO. In the initial fielded version of FLEX, relatively rudimentary alert messages will be supported.[7] Later versions of FLEX will implement more sophisticated alert messages.

Table 6.1

FLEX Marquee Icons

Meaning	Icon
Flight Marker (take-off and landing times)	|
Target	Δ
Refueling (refueling mission/being refueled)	⬭/ ○
Rendezvous	R
Crossing	X
Orbiting Aircraft (time on orbit to time off orbit)	⬭
Flight Indicator (mission is flying)	▬▬
Turnaround Time	▪ ▪ ▪ ▪ ▪ ▪ ▪

SOURCE: LaBatt, n.d.

---

[5]Mission A is considered bundled to mission B if A relies on B to accomplish its mission (e.g., a fighter is bundled to a tanker that it receives fuel from and a jammer it receives cover from).

[6]The compilation of all bundled aircraft in a large strike package in a sizable ATO can involve a large number of database queries, is computationally intensive, and can degrade overall system performance with current CTAPS hardware.

[7]The initial version of FLEX will implement the automated alert messages CAFMS generates when mission status reports conflict with the published ATO. For example, when an air base closure message is received by CAFMS, it issues an alert message for each mission that was originally scheduled to fly from that air base.

Implicit in these functions is the capability to detect and predict deviations from the published ATO based upon status reports. FLEX will include automated ATO monitoring and tracking capabilities to achieve this functionality. If a mission is changed and impacts later missions in the ATO, FLEX will determine which subsequent missions are affected. Finally, when deviations from the ATO are detected by FLEX, the system will provide replanning options to the user to eliminate problems caused by the deviations.

ATO Replanning Tools

FLEX replanning tools will be adapted from tools already present in APS. In particular, the Autoplanner used in APS will be incorporated into FLEX. FLEX replanning tools will automatically recognize whether replanning can be done in the following environments:

- Static (when changes occur just after the ATO is published).

- Non-stressed, dynamic (when sufficient time is available for replanning, CTO dissemination, and detailed mission planning).

- Time stressed dynamic (when insufficient time is available for replanning, and replanning time must be traded against time for mission execution).

The FLEX autoplanner will be able to generate and evaluate replanning options when queried. It will also determine the scope of replanning sessions to avoid wasting time and system resources on unnecessary replanning activities. To do this the system will be able to recognize the differences between

- minor plan changes (which affect few unrelated missions, packages, or bundles)

- medium plan changes (which affect several interdependent missions, packages, or bundles—the total number being affected less than 15 percent of preplanned missions in the ATO)

- major plan changes (which may affect more than 15 percent of preplanned missions in the ATO).

Database Management

FLEX may have some database management capabilities many CTAPS mission application modules currently lack, e.g., the ability to automatically exchange data with external databases without user intervention. The external applications FLEX will exchange data with (manually or automatically) and the types of data exchanged are shown in Table 6.2.

At the present time it is not known whether FLEX will automatically exchange data with all the applications listed in the table, and in particular whether FLEX target and OB databases will be automatically linked with those maintained by RAAP and ICM. It is also not known which applications will be able to automatically request and receive data from FLEX.

Table 6.2

FLEX Data Exchange Capabilities

Application	Data Type
ADS	ACO
APS	ATO, ABP, Scenario Data
CAFWSP	Uniform Gridded Data Fields
CAFMS	Scenario and Mission Data
ICM	OBs
JDSS	Requests, Statistics
RAAP	TNLs, CTOs

SOURCE: Griffiss AFB, 1994b.

The generic FLEX database interface model is shown in Figure 6.2. FLEX should be capable of automatically interfacing with databases that are common to the CTAPS architecture. The FLEX statement of work states, "The specific method for interfacing with external systems shall evolve during the interface development process."[8] The FLEX database interface design should become more concrete as the program progresses.

Potential Future Capabilities

A number of potential FLEX capabilities have been proposed. Two such proposals are described below. To date neither development option has been funded. In addition to these proposals, FLEX may also be modified to serve as a decision aid for the theater missile defense (TMD) mission. Discussion of this new potential capability is beyond the scope of the present investigation.

Advanced Graphical User Interface. FLEX is being designed with a lot of user involvement. Users have expressed interest in an improved user interface (UI). The Marquee icon-based display is well liked by operators who have tested prototypes. However, at present its design icons are static entities and cannot be moved or manipulated by the user. In fact, only very limited types of replanning can be done in

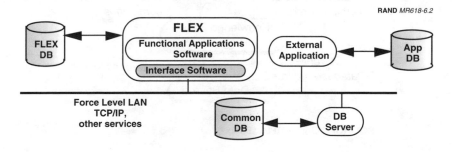

SOURCE: Griffiss AFB, 1994b.

Figure 6.2—FLEX Database Interfaces

---

[8]Griffiss AFB, 1994b.

the current Marquee display, and no graphical display is available in the replanning section of FLEX.[9] Database queries and manual changes to FLEX databases must be made by making text entries on a worksheet or by using dialog buttons.

Operators have expressed a need for a more powerful GUI in which icons could be manipulated or moved to carry out replanning activities or to change FLEX databases. For example, package or target assignments of aircraft could be changed by dragging and dropping icons onto an appropriate part of the screen where the new package rendezvous point or target is located. One can imagine new types of GUI screens which would display relationships between air missions, and in which such relationships could be changed by manipulating icons on the screen.

Software engineers are investigating whether advanced GUIs can be developed to support replanning activities. There are however some difficult technology trades that must be made to determine whether and how to modify the current FLEX application architecture. Presently, the FLEX UI communicates to FLEX databases by means of an intermediate knowledge base program (KBP), as indicated in Figure 6.3. Text-based commands are sent by the UI to the KBP which interprets these commands and issues appropriate structural query language (SQL) questions to the database. The database replies by sending database records or SQL replies to the KBP, which in turn compiles these into the format needed by the UI. The UI then displays these data as lists, or it interprets and displays them as icons.

To display an icon a complex series of messages is transmitted between the database, KBP, and UI. If cursor and icon movements were to be interpreted by the UI as a series of SQL commands (to establish relationships between database entries) a large number of messages could be generated which could seriously degrade overall system responsiveness. More work needs to be done to ensure that not too great a performance penalty has to be paid to provide a GUI-based replanning capability.

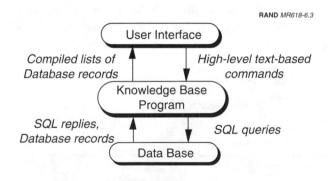

RAND *MR618-6.3*

SOURCE: Griffiss AFB, 1994b.

Figure 6.3—Current FLEX User Interface Design

---

[9]Tabular worksheets are used for replanning in FLEX as they are in APS.

Extension to Airborne Command Elements (ACEs). FLEX could also be tailored as a stand-alone application for use on board AWACS, Joint STARS, or the ABCCC. Currently, operators on these aircraft use paper or electronic copies of the ATO (the latter stored on notebook PCs). During combat operations it may be difficult to manually search the ATO in a timely way for required coordination information when only a paper copy or a PC "flat file" is available. It can be even more difficult for an ACE to keep track of last minute changes to the published ATO when only a paper ATO and pencil are available. If FLEX (i.e., a CTAPS terminal) were deployed on a C2 aircraft it could provide the following capabilities:

- automatic ATO filtering, sorting, and search functions

- status reporting[10]

- mission monitoring

- situation awareness (with planned ATO animation tool)

- replanning.

Much more is required however, if FLEX is to be exploited fully by commanders on airborne C2 aircraft. The ATO monitoring, deviation detection, and replanning capabilities of FLEX will enable the JFACC and his AOC staff to make a relatively large number of target changes in a dynamic battlefield environment. However, if a large number of changes are made during ATO execution and the ATO is to remain useful to ACEs, these changes must be communicated in real time to these aircraft. Hence it is important to have ground-to-air and air-to-air datalinks available to transmit ATO changes to ACEs tasked to carry out real-time battle management activities.

BATTLEFIELD SITUATION DISPLAY

The second major addition to CTAPS 6.0 is BSD. It will permit the JFACC and AOC personnel to view planned missions, near real-time air picture information received from external sources, and intelligence, logistics, and weather data all on one common large screen display.

BSD is illustrated in Figure 6.4. The data displayed can be from sources internal and external to CTAPS. Information displayed will be derived from the ACO, and from CTAPS OB, logistics, weather, and ATO databases. BSD will display air track data received from TADILs, and signals intelligence, missile warning, and other messages received by Constant Source. It may also be capable of receiving message traffic and database updates from other joint or service automated information systems, including maritime tracks from the Navy JOTS system, and ground tracks from the Army and Marine Corps Command and Control systems.

---

[10] Currently, ACE status reports are filed verbally with the AOC.

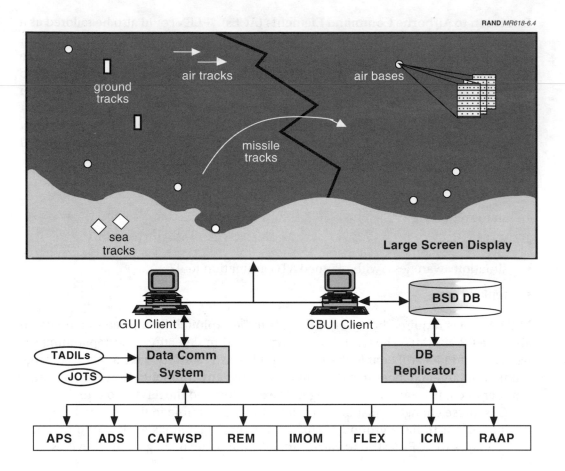

SOURCE: CTAPS Bidders Library, n.d.

Figure 6.4—BSD Database and Message Interfaces

There are three key components to the BSD system. As shown in the figure, BSD will be equipped with its own database, a database replicator, and a Data Communications System (DCS). The DCS will translate messages and will reformat air, maritime, missile and ground track messages received from external sources. These track messages will then be transferred to a GUI client machine which will display them on a local screen or on a large screen display. All graphical data will be handled by the BSD GUI client machine including map and weather data.

The BSD database replicator will take database items from CTAPS applications and integrate them into a single BSD database. Items in this database will be formatted for display by the BSD character-based user interface (CBUI) client machine. The CBUI client will display these data on a local or a common large screen display.

BSD will be responsive only if it can automatically interface with the individual databases of other CTAPS applications and with common CTAPS databases. At the present time it is not known to RAND whether BSD will be able to automatically extract data from all the applications shown in the figure. This problem resembles the one related to FLEX discussed above. BSD should be capable of automatically interfacing with the CTAPS master OB, target, and published ATO databases. Otherwise, a great deal of operator time may be needed to manually enter new OB data or target updates into the database used to drive the BSD display.

A second BSD interface issue is the complexity and feasibility of BSD interfaces to external joint systems. Serious difficulties have been encountered in trying to integrate C4I systems into a single Joint Task Force (JTF)-level system. Recent efforts to solve this problem have led to the Joint Universal Data Interpreter (JUDI) and the GCCS programs, and to the formation of a Joint DoD Standard Data Model. However, in some of these programs, progress has been slow.

In the near term CTAPS connectivity with external joint systems will mostly be in the form of USMTF traffic. However, several different USMTFs are used by the services. A translator capable of processing about 15 different USMTFs would be needed to provide a full situation-assessment display, like that envisioned with BSD or GCCS.[11] A large number of messages may be received by BSD with many in incompatible formats. If these messages are to be automatically processed and displayed, a complex USMTF processing system would be required. USMTF standards are also dynamic. Anywhere from 300 to 600 message changes are approved each year. These changes would require similar changes be made in the associated message processing systems. *Given the technical difficulty associated with Joint C4I system USMTF interoperability, it may be appropriate to initially limit the number of external joint system interfaces in CTAPS and BSD.*

If BSD can responsively display battlefield data, it will provide a significant improvement in JFACC and AOC staff situation awareness. The addition of a large screen display to the AOC CTAPS architecture will allow the JFACC and his staff to collectively estimate how ATO execution is proceeding by comparing near real-time air picture information with data from the published ATO. These capabilities will significantly enhance the collaborative decisionmaking environment in the AOC. What is needed by high-level commanders and combat operators is an accurate composite air picture. Various technology solutions to this data fusion problem have been investigated, but the technical challenges associated with it have yet to be resolved. The BSD represents an interim solution to this long-term problem.[12]

The secondary air picture information received by BSD may require careful interpretation. Air track data received by BSD may be subject to unknown time delays due to the surveillance or communications systems used. Because of these delays it may be difficult to interpret and use such air picture data. If the air picture is too "old" it may not be useful to high-level commanders. A similar interpretation problem may occur with simulated air pictures produced from ATO data and models of aircraft route profiles. *The utility of BSD and these types of information displays should be evalu-*

---

[11]Kameny et al., 1994.

[12]BSD will not produce such a picture but will present the planned air picture (from FLEX) and the actual air picture (from AWACS or other TACS elements) side-by-side.

ated in live flight exercises, ideallly on test ranges and not just in command post exercises.

INTEGRATION OF INTELLIGENCE SUPPORT SYSTEMS

Version 6.0 of CTAPS is scheduled to be released in FY 96. At that time new versions of CIS and JDISS mission applications will be integrated into CTAPS. We examine how these systems could be integrated into CTAPS 6.0 and some of the issues associated with integration.

The planned CIS architecture will be composed of the same complement of mission applications: ICM, RAAP, SB, and CS. The Air Force plans to integrate these (they currently have limited interoperability with each other) into a set of interoperable mission applications.

System Connectivity

The planned connectivity between these systems—to the extent that it can be currently specified by RAND—is shown in Figure 6.5. Where automatic database exchanges will definitely be possible these links are shown in black. There are links where automatic database exchange may be possible, but the degree of interoperability that will be possible was not known to RAND at the time of publication. Where uncertainty remains about automatic data exchange these links are shown in gray.

Because of the evolutionary acquisition strategy used to develop these systems, and continued evolution of the underlying COTS relational database management sys-

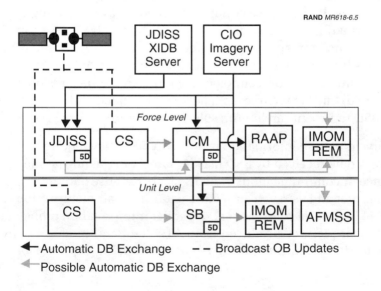

Figure 6.5—Planned Integration of Force- and Unit-Level
Intelligence Systems

tems, it is difficult to determine in advance whether automatic links can be established between databases. Clearly, the goal should be to develop automatic database exchange capabilities for all the connectivity paths shown in the figure. However, because JDISS is being developed independently of CTAPS and CIS it will naturally be more difficult to ensure automatic database exchange with this system. In the case of JDISS and other external systems, lower level communications and message interoperability should be ensured however. It may be appropriate for the Air Force to encourage DIA to take the same approach as the CIO—develop a JDISS client tool software that can be incorporated into CTAPS as a mission application. This will make it easier to achieve automatic database connectivity between systems, although it will mean that the DIA will have to port JDISS software to the different open system hardware platforms used by each of the services.

Two important advances will be made when the systems shown in Figure 6.5 become operational. First, the integration of CS with the other CIS and CTAPS mission applications will permit OB updates and threat warning messages to be incorporated quickly into other CIS OB databases—perhaps in near-real time if automatic database exchange is possible. This promises to greatly improve the responsiveness of all intelligence systems used at the force and unit levels.

The second advance will the ability to receive and process imagery data in a common open computer system environment, both at the force and unit levels. This will be made possible by integrating Air Force systems with those being developed by the CIO in the Warrior Vision program.[13] The CIO is developing a uniform set of secondary and primary digital imagery standards and imagery file compression standards and is modifying 5D imagery workstation software so it can run as a mission application on ICM and SB workstations and on other joint systems. The CIO will establish a number of regional imagery servers that will be available on-line as File Transfer Protocol (FTP) or Mosaic-like sites on the Defense Information Systems Network (DISN). The 5D image retrieval client tool will be capable of requesting imagery products, remotely searching CIO imagery databases on regional CIO servers, downloading soft copy imagery files, and transmitting imagery collection requests to the CIO.[14]

Because of budget cuts, the CIO will no longer provide hard copy imagery products. All imagery will be provided in soft copy form. What are not shown in the figure are the LAN and WAN networks that will provide connectivity between all these systems. Air Force LANs at the wing and unit levels will be used in addition to larger joint WANs that will be overlaid on tactical communications networks and the DISN. Because of increasing remote database access requirements, for example between CIS and SB, and demand for soft copy imagery products, overall total demand for digital communications will increase significantly in a major contingency.

---

[13]This program is the "imagery" part of the JCS GCCS program.

[14]The CIO will evidently act as a broker of such requests, and in some cases the CIO has promised specific turnaround times on collection requests.

The addition of imagery file traffic generated by the 5D client tool will greatly increase communications traffic on force- and unit-level LANs. In its current configuration the CTAPS LAN is sometimes fully loaded with database-related traffic passed between servers and clients, so imagery files may not be effectively transported on current LANs within the AOC. The new imagery dissemination architecture illustrated in Figure 6.5 may require a new force-level LAN to be implemented.

A similar problem may exist at the unit level. The CIO has tentatively identified a requirement for a full T1 communications link to each wing to support existing communications needs and emerging needs for imagery data.[15] There is uncertainty about this requirement however. Originally, the Air Force was to establish an Air Force–wide WCCS LAN program to support computer communications at the wing and squadron levels. However, because of funding shortfalls and a possible duplication of effort with wing in-garrison programs, the WCCS LAN program was canceled. Consequently, emerging digital imagery transport requirements should be folded into local wing LAN programs.

In the past the most difficult obstacle in realizing a vision like that shown in the figure was communications interoperability. Different communications systems, network standards, and data file and message standards were used. Now that joint agencies and the services are adopting open computer systems and Internet-compatible networking systems (routers and servers), the biggest obstacle may be sufficient communications capacity to support all the new information services that will become available on the DISN and increased information traffic on tactical LANs and WANs. Significant requirements study and program coordination will be needed by the Air Force and joint agencies to ensure that sufficient capacity will be available.

The specification and incorporation of digital imagery and remote database communications capacity requirements is a major architectural issue that should be addressed in a number of Air Force and joint programs.

INTEGRATION OF THE AIR CAMPAIGN PLANNING TOOL

A second important CTAPS system configuration issue concerns how ACPT will be integrated into CTAPS, what functions it will perform or support, and which CTAPS mission applications it will interface with. There appear to be two options for incorporating ACPT into the overall CTAPS architecture. Both of these options are predicated on the assumption that ACPT will be used as an automated support tool for target selection and MAP production. If ACPT is used only for target selection purposes, the current floppy disk interface between ACPT and CTAPS is probably adequate once a translation program is developed to speed the incorporation of these data into CTAPS databases.

In the first option (Figure 6.6), ACPT would furnish a prioritized strategic target list and the MAP to CTAPS mission applications, but CTAPS and ACPT databases would

---

[15]A T1 link is a high-capacity communications link that provides a total capacity of 1.54 megabytes per second (Mbps).

remain distinct and separate entities.[16] A depiction of the ACPT, CTAPS application connectivity, and the data exchanged for this option are shown in Figure 6.6. Because ACPT would be used for MAP production, it would have to access force status and logistics databases (for example to obtain munitions and fuel availability data) in CTAPS. This may require communication with databases managed by other mission applications not shown in the figure.

It may be possible to implement the connectivity shown in the figure even if ACPT still operates at a higher security level than CTAPS. Multilevel security (MLS) interfaces would be required between ACPT and ICM, RAAP, APS and perhaps other CTAPS logistics databases. Two of these MLS interfaces, for MAP transmission to APS, and for the transmission of logistics data to ACPT, could be relatively simple to implement, since they would support only one way data flows in highly structured database formats.   In this scheme ACPT would maintain a separate targeting database at a higher classification level. Secret-level targeting databases would be maintained by RAAP, ICM, and APS, but they would be provided sanitized updates from ACPT. Other target-related information could flow in the opposite direction as

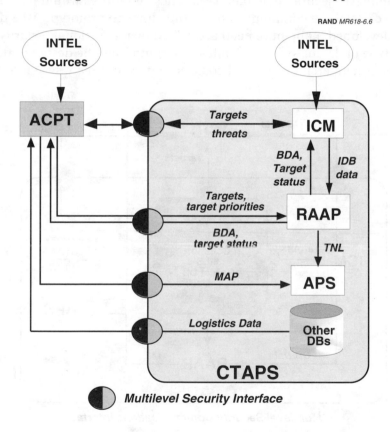

Figure 6.6—ACPT Integration—Option 1

---

[16]The MAP furnished to CTAPS may be somewhat notional or not precisely deconflicted (see the discussion in Chapter Five for details).

shown in the figure. Of course, it would be simpler to develop an MLS interface for linking ACPT and CTAPS target databases if a single CTAPS mission application target database was made the master CTAPS target database, or if a common CTAPS target database was developed that all CTAPS applications shared.

In the first option, detailed target development and weaponeering would still be performed by RAAP, and OB database management by ICM.

The second ACPT-CTAPS integration option would entail a much greater degree of integration between ACPT and CTAPS. In this option, ACPT would also be used for prioritizing targets and as an automated MAP production tool, but CTAPS and ACPT target databases would be merged into a single multilevel secure database. One possible connectivity application and database profile is illustrated in Figure 6.7 for this option.

If ACPT were to remain at a higher classification level than CTAPS, the type of connectivity envisioned in option two would require sophisticated MLS interfaces to a common multilevel secure target database. These interfaces would have to filter a wider range of database communications. And although prototype MLS databases have been developed, none have been accredited by the National Security Agency. Use of this type of database could significantly complicate the design of the CTAPS client/server system architecture and could introduce untimely system integration

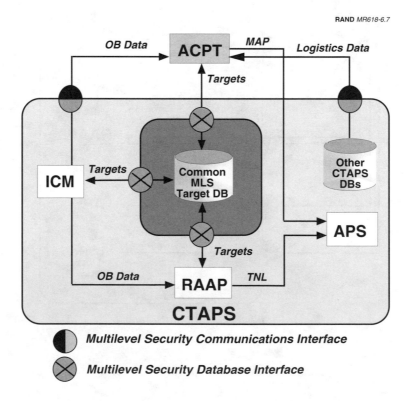

Figure 6.7—ACPT Integration—Option 2

delays when new computer or network hardware is introduced into the CTAPS architecture. Because of the frequent hardware and software upgrades that are characteristic of open system COTS products, the use of a MLS database within the CTAPS architecture could substantially increase the chance of development delays and complicate the rapid prototyping environment that has been a characteristic of the CTAPS program.

If this option were selected, it would be much easier to implement if ACPT were downgraded to a secret-level system. This would eliminate the need for a special multilevel secure database. Significant efforts have been undertaken at the national level in efforts to downgrade many intelligence products to the secret level. This will make intelligence products more readily available to the war fighter and will make it easier to integrate intelligence systems into tactical C4I systems. What is not known to RAND at the present time, is what significant ACPT functionality would be sacrificed in downgrading ACPT to the secret level. The transparency of the links that ACPT establishes between high-level goals and specific air campaign plans or targeting strategies may be lost or obscured by sanitizing high-level guidance or intelligence information used in the STT process.

CTAPS AIR SUPPORT OPERATIONS CENTER AUTOMATION

The Air Support Operations Center (ASOC) is the key interface between the Army Corps and the AOC. The ASOC informs the AOC of Army CAS and battlefield air interdiction requirements, assists in corps air campaign planning, informs the corps and subordinate Tactical Air Control Parties (TACPs) of current and planned air operations, and operates the Air Force Request Net (AFRN). The primary mechanism usually used to coordinate Army air support requests with the AOC is the ATO. It is therefore important that the ASOC and TACPs have timely access to the ATO, and to ATO change messages.

Air Support C2 During Operation Desert Storm

ATO interoperability problems were experienced by the Army during ODS. The ASOC was equipped with a very limited number of remote CAFMS terminals. Subordinate TACPs were not equipped with CAFMS and had no reliable means other than courier to receive the ATO. During ODS the ASOC was located at corps headquarters and was typically far removed from the forward locations where TACPs were located. In many cases, missions had already flown before TACPs received the ATO. In general during ODS, the TACPs did not have timely access to the ATO.[17]

Even if the TACPs had been equipped with CAFMS, they still would not have had timely access to the ATO because they lacked high capacity data communications to the ASOC. The AFRN, which links the ASOC with TACPs, has relied upon high-frequency or very high frequency line-of-sight (LOS) radio links. VHF radios have

---

[17] Whitehurst, 1993.

limited range and neither type of radio can easily provide high data rate communications.

Lengthy delays were encountered at the ASOC in downloading the ATO from the TACC (AOC). It took a remote CAFMS terminal about one minute to access a single "page" of the ATO (about 20 lines of data). Because the Desert Storm ATO was frequently over 800 pages long, it often took an operator using a remote terminal over 10 hours to download the entire ATO. These delays occurred because CAFMS used 1200 baud modems and an outdated non-packetized communications protocol to connect remote terminals to the central ATO database. Similar delays were also experienced at Air Force WOCs.

There were other reasons why CAFMS could not provide responsive remote access to the ATO. Each CAFMS "mainframe" computer could support a *maximum* of only 12 terminals. In ODS five CAFMS "mainframe" computers were deployed and networked together along with 54 remote terminals. All available CAFMS terminal and "mainframes" were used. There simply were not enough remote terminals for all users who needed one. Users who did have access to the CAFMS network overloaded the system, and their demands for database access from within the AOC or from remote locations could not be met.

This type of difficulty may also occur in the future with CTAPS although for different technical reasons. There is no hard and fast limit on the number of CTAPS terminals that can be connected to a LAN or a WAN. However, the number of remote CTAPS terminals planned for deployment to various joint C2 centers has grown significantly. CTAPS remote access capabilities should increase correspondingly. Otherwise, CTAPS databases or the CTAPS WAN may prove to be bottlenecks in future operations.

ASOC Automation Program

The ATO dissemination problems encountered during ODS have prompted development of a new ASOC automation system that will be a part of the CTAPS architecture. Maximum use will be made of existing CTAPS mission applications. The ASOC system will be based on the same underlying COTS and GOTS software architecture and will use the same CTAPS communications interfaces for ATO and CTO dissemination. It is expected that CAFMS, APS, and possibly FLEX mission application modules will provide the core capability of this system. The system will receive the ATO, process immediate and preplanned air support requests, provide automation support to related Army Corps air support planning, and support ASOC coordination efforts with the AOC and Battlefield Coordination Element (BCE) over the CTAPS WAN.

The CTAPS ASOC system will have a WCCS interface to communicate air support requests to the unit level. Connectivity to the wing level and AOC will be provided by extension of the CTAPS WAN. This network will be overlaid on existing Air Force and Army tactical digital communications links, including the Army MSE Tactical Packet Network (TPN).

Digital communications interfaces for targeting and retasking messages may also be integrated into CTAPS ASOC systems. Forward observers equipped with a Digital Communications Terminal (DCT) will relay targeting and retasking messages to in-flight aircraft equipped with the Improved Data Modem (IDM) or an interoperable equivalent. The improved version of the AFRN will have some sort of DCT communications interface to provide an automatic data transfer capability to forward observers.

A prototype ASOC is under development and being incrementally tested by the Ninth Air Force at Shaw Air Force Base.[18]

---

[18]Because of limited resources and time available to conduct this study, we have not been able to examine the ASOC Automation program in detail.

Chapter Seven

CTAPS IMPLEMENTATION OF THE ATO PROCESS

In the last two chapters we reviewed the CTAPS 5.0x and 6.0 architectures, the functions of CTAPS mission applications, and those of other related automated support systems used in the air campaign planning process. In this chapter we examine how information flows between CTAPS mission applications modules and external systems during the ATO production process.

First, we consider how information flows during ATO production in CTAPS 5.0x. We identify bottlenecks in the process caused by limitations in this version of CTAPS. Then we review the potential information flows and database features of CTAPS 6.0 and examine how these features may improve or impair the ATO production process.

CTAPS 5.0x ATO PROCESS ARCHITECTURE

A graphical representation of the CTAPS 5.0x ATO process architecture is depicted in Figure 7.1. CTAPS mission application modules and related external systems are indicated respectively by boxes and shaded ovals. Arrows indicate the direction of information flow between systems and modules. Solid lines indicate automated information links such as database file transfers, and dashed lines indicate manual information transfers—information that must be typed into the system or "cut and pasted" from one application to another by an operator.

Thick arrows signify the major information flows required in the air campaign planning process. We examine the latter set of processes in detail below.

The divided background of Figure 7.1 represents the four divisions of an AOC. Mission applications that lie on the boundary of two divisions are used in both divisions. Two copies of CAFMS and other mission application modules are shown in the figure. This indicates that multiple copies of an application are used in two or more divisions of the AOC, or that different databases of the same application are used in two or more AOC divisions (e.g., while CPD personnel work on ATO B, COD personnel use the preceding published ATO, or ATO A).[1]

---

[1]The figure does not accurately reflect the number of application copies or databases used in the planning process.

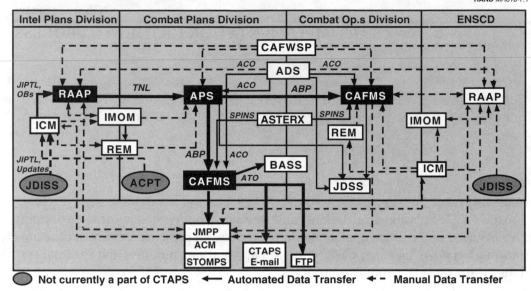

Figure 7.1—CTAPS 5.0x Information Flow

The bottom section of the figure represents the CTAPS suite of communications-related applications. Mission applications that can send or receive message traffic from external units have dashed lines leading from them to ACM and JMPP.[2] All AUTODIN message traffic exits or enters through the STOMPS application as shown. As indicated in the figure the ATO can be disseminated by AUTODIN, CTAPS E-mail, or the Internet FTP program.

ATO Cycle Implementation

The ATO cycle starts when the most recent OB and targeting information received from external sources is incorporated into CTAPS OB and target databases. As indicated in the figure this information can be received through JDISS.[3] Strategic targets are usually prioritized by the JFACC and his AOC targeting staff in the CPD. Strategic target prioritization is usually done manually. However with the introduction of ACPT, strategic targets can be prioritized automatically and then manually forwarded to ICM as shown in the figure.[4]

In CTAPS 5.0x, target and OB file transfers from JDISS and ICM have to be performed manually. Application developers for the two systems have employed different COTS database products and configured these databases differently. A requirement for

---

[2]For simplicity, not all these links are shown in Figure 7.1.

[3]As the reader may recall the JIPTL typically contains a list of prioritized tactical ground targets specified by the Land Force Component Commander and his joint service representatives.

[4]Information can be transferred from ACPT to CTAPS via floppy disk, but the data must be reformatted and parsed for inclusion into CTAPS databases.

automatic electronic file transfer between JDISS and ICM has been identified by operators in recent exercises.

Once all candidate targets have been received by ICM they can be transmitted electronically as a *single* database from ICM to RAAP. New targets or target updates cannot be automatically transmitted between ICM and RAAP however, and transmission of a moderately sized target data base between these two applications can take from up to four hours to complete with the current CTAPS hardware suite.[5] Consequently, transmission of the candidate target list to RAAP can be done at most once a day with current-generation workstations.

In RAAP, targets are weaponeered and put in the TNL. The TNL is then transmitted to APS. To transfer the TNL, the entire target database must be passed, which may also be time consuming depending upon the size of the database. Problems were encountered in passing the TNL to APS in an early version of CTAPS 5.0. The problem was fixed in version 5.06. However, the complex software environment that caused it remains.[6]

Once the TNL and the MAP (currently produced manually as paper worksheets if ACPT is not used) are input into APS, planners use APS to build the ABP. After the ABP is finished it is transferred to CAFMS as indicated in Figure 7.1. The transfer of the ABP to CAFMS also sometimes failed in exercises in FY 94 when an early version of CTAPS 5.0x was used. This problem stemmed from the inability of a translation program to translate certain elements of the ABP message. This problem was fixed in CTAPS version 5.06. However, this difficulty points out a larger issue that must be addressed as the CTAPS architecture expands and evolves. *Careful configuration control over data element and data storage format standards must be maintained across databases, applications, and communications programs to ensure interoperability problems do not creep into the architecture.*

CAFMS is used to collate the ABP, SPINS, and ACO to form the ATO. Once collation has been completed in the CPD, the ATO can be transmitted by AUTODIN, by CTAPS Email, or it can be saved as a "flat file" and transferred to terminals on the network using FTP. As shown in the figure, the ATO is also transferred to the BCE Automated Support System (BASS) used in the Army liaison cell in the AOC. Since the BCE is a unit that is established only in wartime, this transfer probably has not been demonstrated in exercises. The Naval Liaison Element has no special purpose ATO system. Increasingly, however, naval personnel, having become familiar with CTAPS, can access the ATO directly using CAFMS.

---

[5]Conversations with CID personnel during Exercise Tempo Brave 94-1. This difficulty is due partly to hardware performance limitations. The current version of ICM, version 1.0, only runs on relatively slow SUN SPARC 2 workstations.

[6]The problem stemmed from a naming convention incompatibility in RAAP when multiple weaponeering options were attached to targets and from the use of non-standard SQL queries. RAAP and APS use Sybase and Oracle relational database management systems (RDBMSs), respectively. Sybase and Oracle are competitors and use different non-standard SQL extensions, making RDBMS interoperability difficult. CTAPS database interoperability would be easier to establish if only one RDBMS were used.

In CTAPS 5.0x, CAFMS is also used by COD personnel as a database management tool. It receives database updates from the wing level via AUTODIN message traffic. Changes to the ATO are recorded in the CAFMS ATO database maintained in the COD. ATO change messages are generated in CAFMS and transmitted via AUTODIN to the wing level. Other ATO changes, such as the retaskings of airborne aircraft, are recorded in the COD CAFMS database and transmitted verbally to airborne C2 aircraft.

CTAPS 5.0x Serial Planning Subprocesses

Having reviewed how target information flows between CTAPS 5.0x mission applications, we now examine how these applications support particular subprocesses in the ATO production process and determine which subprocesses are performed in parallel and which are performed serially. Limitations in the way CTAPS databases are currently connected restrict how the current ATO production process can be shortened and the degree to which subprocesses can be performed in parallel.

The ATO production process is implemented as a set of largely serial subprocesses in the CTAPS 5.0x. These sequential subprocesses, CTAPS mission applications, the external systems involved, and the information flows between applications and systems are illustrated in Figure 7.2. In the order in which they are performed, these subprocesses are target development, MAP production, weaponeering, ATO production (ABP, ACO, and SPINS), and ATO collation and dissemination. The particular CTAPS mission applications and systems used to support each subprocess are shown in the figure.[7]

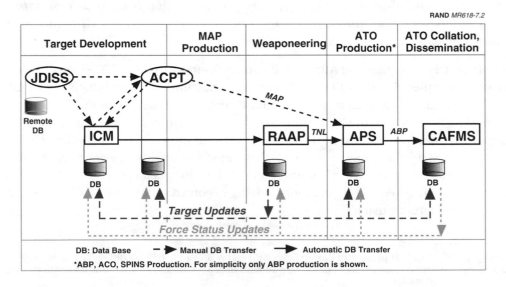

Figure 7.2—CTAPS 5.0x Mission Applications and ATO Production Subprocesses

---

[7]For simplicity, in the figure RAAP is placed under one subprocess—weaponeering. In actuality, it is also used for target development.

The first subprocess of CTAPS ATO production is target development. In it, targeting information is received from external sources.[8] We include the selection and prioritization of targets as part of the target development subprocess. The strategic target list can be selected and prioritized by using ACPT, if it is available. Next, the MAP is produced. MAP production can be done manually or by using ACPT to produce a notional MAP (as shown in the figure). Next, weaponeering is performed. These first three subprocesses in the cycle are shown being done in serial in the figure. This is the order in which they were performed with CTAPS in recent exercises and the way the ATO production process is described in current doctrine.

After the TNL and MAP are transferred to APS, detailed ATO production is carried out. For simplicity, in the figure only APS and the ABP production subprocesses are shown. The SPINS and ACO are produced in parallel subprocesses using different applications. We focus on the ABP because its production is by far the most difficult and time consuming. After the ABP is finished, the ATO is collated and disseminated in the last subprocess in the cycle.

Excluding transfers of MAP data, most database transfers between subprocesses can be performed electronically in CTAPS over the LAN. However, in each case the entire database must be transferred. Because it can take considerable time to transfer a database, each such transfer can be done only once in the ATO cycle. This limits which subprocesses can be done in parallel because, as shown in the figure, the database outputs of one subprocess are required as inputs for the next subprocess. Because database transfers link most subprocesses, most subprocesses can only be executed serially in the ATO cycle.

The only way to get around this limitation is to make updates simultaneously to each database affected by an update or change. However, updates to target or force status databases must be made manually in CTAPS 5.0x. The same is true in the case of OB databases. The limited database interoperability of the current CTAPS architecture makes it difficult to implement the major subprocesses of ATO production in parallel fashion.

Target Databases

As Figure 7.2 illustrates, if a target is added or changed during the process, up to four separate databases have to be updated to maintain database accuracy throughout the system. If ACPT is employed in the process, up to five separate databases may have to be updated with the same information depending upon when in the ATO production process the change or update is made. Because such changes have to be entered manually, handwritten, verbal, or CTAPS E-mail messages must be transmitted to the relevant duty stations to alert database managers of targeting changes.

In principle, new targets could be added at any point in the process and each database updated manually. However, this may disrupt the planning process—especially if many changes are made at random—because operators will have to spend

---

[8] For simplicity, only a JDISS link to external sources is shown in the figure.

valuable time communicating and making changes, and because *in most cases all planning activities involving CTAPS must cease while target databases are updated.*

For these reasons the number of replicated target databases in the current CTAPS architecture may make it difficult to maintain target database integrity during an operation involving a large number of targets. In the CTAPS 5.0x architecture, four target databases must be synchronized to maintain database integrity.

OB Databases

In CTAPS 5.0x, separate OB databases are used by at least five CTAPS mission applications. These applications are shown in Figure 7.3 on the right-hand side of the figure. In practice initial OB databases for each CTAPS mission application are loaded separately in garrison. OB updates are received from external sources, some of which are shown on the left-hand side of the figure. Updates to each CTAPS mission application OB database have to be made manually, as indicated in the figure.

OB data can be forwarded from ICM to RAAP, IMOM, REM, APS, and CAFMS. But none of these databases are linked. ICM can electronically transfer OB data to IMOM or REM, but the entire database must be transferred. This process can also take a significant amount of time depending on how big the database is. OB updates received in the midst of the planning process therefore must be entered manually into each of the individual applications. A significant amount of manual operator activity is required to maintain OB database integrity throughout the CTAPS 5.0x architecture. In the dynamic environment characteristic of air battle planning in an actual operation, it will be difficult to ensure that seven different CTAPS OB databases remain synchronized.

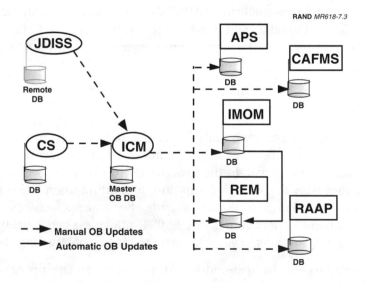

Figure 7.3—CTAPS 5.0x Order of Battle Databases

OB databases are essential to the ATO cycle. Havoc can be wreaked on the planning process if different AOC divisions use different OB databases. This problem occurred during ODS, especially between unit- and force-level organizations. Work arounds were found that improved OB database accuracy. However in the future, the threat may be more onerous and less time may be available for manual OB coordination. If the ATO cycle is shortened, automatic compilation and distribution of a single, accurate, and comprehensive OB database to CTAPS applications and to unit-level organizations will be a critical requirement.

The complicated database interactions between CTAPS mission applications and the external systems shown in the figure stem from the evolutionary acquisition strategy used to develop these programs. Many CTAPS mission applications originally started as separate programs and so were designed to rely solely on their own OB and target databases. CS and ICM also originally started as stand-alone intelligence support systems. CTAPS 5.0x represents the first effort to integrate these disparate development efforts into one system. There are plans to better integrate both ICM and CS into the CTAPS architecture. Improved integration of OB databases should be a priority in future software upgrades of CTAPS and CIS mission applications.

POTENTIAL CTAPS 6.0 ATO PROCESS ARCHITECTURE

The planned CTAPS 6.0 architecture could significantly improve the air campaign planning process. As discussed in the preceding chapter several new mission applications will be added. However, these new applications will present new database integration challenges to CTAPS systems integrators. Below, we examine the implications of these new applications for the air campaign planning process, and the additional database integration measures that may be needed to realize the promised improvements of version 6.0 of CTAPS.

Some of the potential information flows in CTAPS 6.0 are shown in Figure 7.4.[9] New mission applications and new interfaces to external systems are highlighted in gray. Mission applications that are expected to be highly integrated are shown merged together. Where automatic database exchanges are expected to take place solid connecting lines are used. Heavy solid lines indicate the major database transactions expected to occur in the ATO production process. Gray lines indicate information flows that may be automatic or manual, depending upon the level of integration and interoperability achieved in the CTAPS 6.0 database architecture. The figure illustrates the complex set of application interfaces that will still likely be present between intelligence support and target development mission applications.

---

[9]At the time this report was written, definitive data were not available on the precise software configuration for CTAPS 6.0. Shown in Figure 7.4 is the best estimate of CTAPS 6.0 mission applications and information flows based upon data available in the CTAPS Bidders library at Hanscom Air Force Base and from other sources.

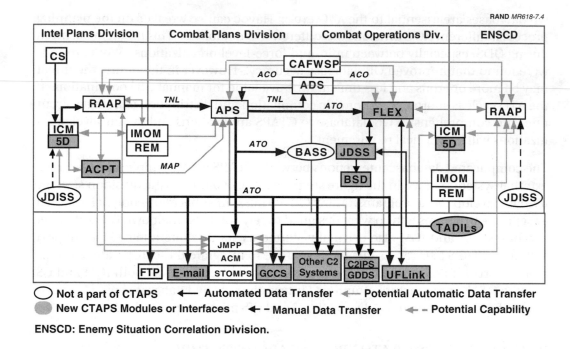

Figure 7.4—Planned CTAPS Architecture

Implementation of the ATO Production Process

As indicated in the figure, the key information flows in the ATO production process in version 6.0 are similar to those in version 5.0x. The information flows in the ATO production process have been simplified somewhat by the removal of CAFMS from the Combat Plans Division. This implies that fewer large database transfers (in this case of the ATO) have to be made during the planning cycle. However more information transfer and database coordination may occur in the target development and weaponeering portions of the planning cycle because of the integration of ACPT into CTAPS.

There are three major structural changes to the CTAPS architecture that may be implemented in version 6.0 which could have a major impact on the ATO production cycle. First, there will be many more ATO dissemination options available because of the increased number of interfaces to external systems. Second, force-level intelligence support systems will be more tightly integrated with one another and with other CTAPS mission applications. And third, ACPT will be integrated into the CTAPS architecture. However, in order for these changes to have a net positive effect on the ATO production process, several database integration issues must be addressed. We review the implications of these changes below.

Version 6.0 of CTAPS will also have several important new capabilities that could significantly improve the ATO execution monitoring capabilities of combat operations personnel in the AOC. These improvements will result from the integration of

FLEX and BSD into the CTAPS architecture. However, in order for these changes to have a net positive effect on ATO execution a number of important database integration issues must also be addressed.

ATO Dissemination Communications.   The first major change in the architecture, improved ATO dissemination communications, is intended to increase joint and intra-service interoperability but could also increase the complexity of CTAPS significantly. The Unit-Force-Level Link (UFLink) will be used for ATO dissemination to wings and the ASOC. The ATO will also be disseminated to the Command and Control Information Processing System (C2IPS) and the Global Deployment Support System (GDSS), which are automated support systems used in the Air Lift Command Center (ALCC). Eventually, automatic updates will also be received by CTAPS from C2IPS when technical issues concerning this transfer are resolved. As before, the ATO can be disseminated to remote locations via AUTODIN, E-mail, or FTP.

The ATO will also be disseminated to GCCS nodes at JTF and CINC command and control centers, and to other C2 systems. As of this writing, RAND could not ascertain which additional systems would have ATO interoperability with CTAPS, although the list appears to be growing with time and probably includes Army command and control systems based on the GCCS Common Operating Environment (COE), the Navy Joint Maritime Command Information System (JMCIS), the USMC Advanced Tactical Air Command Central System (ATACCS), and the U.S. Marine Corps Improved Direct Air Support System (IDASS).

Developing and maintaining interoperable communications links with other service units for timely ATO dissemination may be difficult in the near term in the absence of commonly agreed upon joint communications, computing, and message processing standards. A universal joint format has formally been agreed to for ATO dissemination which will make ATO dissemination much easier; however problems will likely remain until the GCCS COE becomes available and the majority of the command and control systems used by the services (including CTAPS) are ported to the GCCS COE.

Examples of this type of problem are already evident. ATO dissemination problems between Air Force and Navy versions of CTAPS 5.0x applications were observed in recent exercises, even though these systems used exactly the same ATO database formats and were connected by interoperable communications links.[10] The services increasingly are using state-of-the-art COTS workstations in their C4I systems. Consequently, these systems undergo frequent hardware and software upgrades, including changes to their underlying operating system software. Furthermore, many different types of COTS open system products are in use in each of the services and defense agencies. These factors make configuration management of a large number of communications links difficult.

---

[10]CTAPS 5.0x software has been ported to the open system workstations used in Navy C2 systems. However, because of subtle differences in the operating system and related software used in Air Force and Navy C4I systems, interoperability problems were introduced that prevented dissemination of the ATO to Navy ships in some recent exercises. These problems have since been fixed in version 5.06 of CTAPS.

The GCCS COE will potentially eliminate these configuration control problems since it will provide a common computer operating system and common basic supporting applications that will run on many of the different hardware platforms used in DoD. It is being specifically designed to be much more platform independent than past commercial open system products. However, the GCCS COE will probably not be available for version 6.0 of CTAPS.

If all major C2 centers and units have fully interoperable ATO dissemination links that allow direct access to the central ATO database at the AOC, the time needed for ATO dissemination could probably be reduced significantly. This reduction could in turn lead to a reduction in the overall air campaign planning cycle and to more effective air campaign plans that include a larger number of force elements.

Integration of Force-Level CIS Applications. As discussed in Chapter Six, the suite of Air Force intelligence support systems used in the AOC (ICM, CS, and RAAP) will be more tightly integrated in CTAPS 6.0. These systems will allow AOC personnel to access remote intelligence databases and imagery files as well as receive intelligence updates broadcast on military satellite communications channels. Improved integration of these systems will help ensure that planners have the most recent intelligence data possible and may reduce the time needed to prepare key databases used in the ATO production process. Improved integration could also reduce operator workloads if intelligence updates could be automatically incorporated into existing AOC OB and targeting databases.

Because most of these intelligence support systems were originally developed as separate stand-alone systems by independent contractors, they do not all use the same relational database management system software, and they store data in different data element formats. In order to achieve the most effective level of integration of AOC and CTAPS intelligence support systems, these systems should all use the same or compatible relational database management systems and should also store data in the same standard data element formats.

Integration of ACPT. The only major part of the air campaign planning process that CTAPS 5.0x does not provide automation support for is MAP production. Integration of ACPT into CTAPS 6.0 will address this limitation and could significantly reduce the time needed to produce the ATO. However, classification issues and the multilevel security integration approach must be resolved. Two possible approaches were described in Chapter Six. In either approach, automatic links would be established between ACPT databases and selected databases maintained by other CTAPS mission applications. If the multilevel security issues can be resolved, the ACPT target database should be automatically linked to similar databases maintained by RAAP and ICM, or all three applications should employ a common target database.

It is most important however to ensure that the notional MAP produced by ACPT be transferable to APS via the CTAPS LAN. It may not be necessary to modify the ACPT MAP format to match the ATO data element formats used in APS if a translation program can be written to transfer ACPT MAP data into APS.

In terms of ATO production, CTAPS 6.0 may only improve the process incrementally. If the changes in doctrine identified in Chapter Three are not made and the majority of CTAPS databases are not automatically linked together, it is likely the same set of serial subprocesses used in the air campaign planning process in CTAPS 5.0x will also be carried out in series in version 6.0. If this is the case, the ATO production cycle time of 48 hours may not be reduced significantly.

Mission Monitoring and Retasking During ATO Execution

The replacement of CAFMS by FLEX and the introduction of BSD promise to significantly improve the information access and manipulation capabilities of the JFACC and his combat operations staff in the AOC. FLEX, BSD, and the CTAPS ASOC automation program should permit the JFACC to manage and monitor air operations much more closely as they are executed and to divert or recover air assets as problems or opportunities develop during operations. However, the introduction of FLEX and BSD will significantly increase the complexity of CTAPS links to the unit level. More real-time information will enter the system from the wings, from Navy units, and the ASOC. In CTAPS 6.0 most of this real-time reporting will likely be received as AUTODIN message traffic, which will have to be interpreted, parsed, and integrated into CTAPS databases in order to provide the JFACC with a unified and up-to-date picture of the air war.

The real-time ATO monitoring capabilities of FLEX will generate increasing demand for timely status reports from the unit level. However, communications from other service units may introduce message interoperability problems. Different USMTF message formats are used by the different services. A universal joint format has been agreed to for ATO dissemination which will make ATO dissemination easier, but MISREP and general situation awareness messages can come in a variety of different formats. Force status, OB, and mission updates may be processed automatically by CTAPS only if they are received from Air Force units that use remote CTAPS terminals. Messages received from other C2 systems may have to be processed manually. Given the volume of such message traffic, updating CTAPS databases with this information could require a great deal of operator time and effort. We hope these problems will be addressed by the Defense Messaging System (DMS) program; however additional message standardization efforts, especially for situation awareness information, will likely be needed to enable automatic status reporting by CTAPS or other DoD C2 systems like GCCS.

BSD will also enhance the situation awareness of the JFACC and his staff by displaying the real-time air picture received from AWACS or ground-based radars. However, if each major CTAPS application maintains its own OB and target databases, then BSD may have to interface to all or most of these databases. As the reader may recall, BSD will maintain its own database that will replicate data in other CTAPS databases. The more databases BSD must replicate, the more complex its design becomes, the larger its replicated database must be, and the slower its own response time will be to database queries. The performance and design of BSD could benefit the most from a rationalization of CTAPS internal databases.

In summary, CTAPS 6.0 will add several important new capabilities that promise to improve the information access and manipulation capabilities of the JFACC and personnel in the AOC and ASOC. However, these new capabilities will significantly increase operator demands for status updates, OB data, targeting data, and imagery files from the unit level, national sources, and from units in other services. In order for CTAPS to be responsive to the increasingly real-time information demands of AOC personnel, its applications and databases will have to be integrated more effectively than they are in version 5.0x. It is unlikely however that all CTAPS applications and databases will be integrated in this fashion in version 6.0, simply because of the present complexity of the system. An evolutionary approach should be adopted to achieve these goals. Such an approach is discussed in the next chapter.

This in turn will increase the overall demand for data communications capacity and for expanded interoperability with other automated C2 systems. These issues require careful study in larger Air Force and DoD C4I architectural studies.

Chapter Eight

EVOLUTION OF THE CTAPS ARCHITECTURE

The scope of the CTAPS program has expanded considerably since it began. Increased emphasis on air campaign planning, the joint employment of air power, and improved methods of ATO dissemination have led to a greatly broadened scope for the program, which now includes all the services.

Care has to be taken in selecting a development path for a program expanding in this way. To select an appropriate path an assessment is needed of COTS product capabilities that are likely to be available in the near-term marketplace. The rapid rate of advance in computer technology, the marketing "hype" that frequently surrounds new commercial products or their expected availability, and making sure the right COTS products are chosen can make the selection of an appropriate development path difficult. If too risky a path is chosen, performance problems may be experienced that could be much more severe than the "growing pains" experienced in recent exercises and stress tests.

The successful integration of complex software into a distributed computer system like CTAPS can be challenging for reasons that are often not predictable. The overall program development risk for a COTS-based system like CTAPS is not really driven by technology development risk, but rather by "integration" risk—the risk that disparate software applications or networked computer systems cannot be made to operate in a "bug-free" manner. Integration risk for complex systems is proportional to the size of the system and the percentage of the system that must be modified to create the required new functions. The CTAPS 5.06 software build now totals more than two million lines of code. As the program grows further in size, the development risk associated with software integration will inevitably grow—at least with the software development technologies used today.

In this chapter, we examine possible future development paths for CTAPS in conjunction with changes to the air campaign planning process. First, we consider how the ATO production process and CTAPS could be reengineered so that significantly less time will be needed for ATO production. Next, we examine fundamental system design and technologies issues that will come into play in selecting a future development path. We also examine CTAPS joint interoperability issues and their implications. Finally, we examine issues associated with the overall acquisition strategy for the program.

SHORTENING THE ATO CYCLE

CTAPS should evolve in a way that provides the functionality most needed by the operator. One of the long-term goals of air campaign planners has been to reduce the time needed to prepare the ATO. In Chapters Two and Three of this report we examined how the ATO planning cycle could be reduced to a 24 hour period by reengineering the ATO production process.

We identified three elements for a reengineered ATO production process:

- Provide automation support for MAP production.

- Reduce the number of target and OB databases in CTAPS, and automate target and OB database update processes.

- Perform target development and weaponeering in parallel with other ATO cycle processes, and schedule when target changes are transferred to MAP and ATO databases.

Below we examine how CTAPS would have to be modified to support a reengineered 24 hour long ATO production cycle.

AUTOMATED MAP PRODUCTION

Near-Term Approach

Automating the MAP production process could significantly reduce the time needed to produce the ATO. In Desert Storm about 11 out of the 48 hours of the planning cycle were devoted to MAP production. It should be possible to cut this time at least in half by using ACPT. ACPT actually provides air campaign planners with important new capabilities that are not possible to duplicate by manual planning even with a large planning staff—the ability to quickly build several notional MAPS and to compare their relative performance in meeting high-level strategic or operational goals. Using ACPT, it may be possible to compare five or six notional MAPS, select one of those for refinement, and to refine the selected MAP, all within four or five hours.

However, to use ACPT effectively with CTAPS, these two systems must be linked electronically. It should be possible to transfer the MAP data produced by ACPT automatically into CTAPS databases (in this case an APS database). Otherwise, a significant amount of time and fair number of planners will probably be needed to enter the data manually into CTAPS. Similarly, it should be possible to automatically transfer targeting and OB data from ICM into ACPT, to ensure that ACPT and CTAPS applications have access to the same target and OB information. As discussed earlier in this report, this does not necessarily mean ACPT and CTAPS have to use a common MLS relational database management system to share targeting information. Targeting information can be shared using carefully formatted target update, or new target messages. However, if ACPT is to operate at a higher classification level than CTAPS, as it probably should, then some type of MLS messaging interface will be needed between the two systems.

For this reason, and because of the unique database structures used in ACPT, it is unlikely that ACPT can be converted into another CTAPS mission application in the near term. However, it can still provide a significant new and badly needed capability to air campaign planners if a minimal set of electronic links are established between ACPT and CTAPS.

Far-Term Approach

Observation of the MAP production process in recent exercises suggests that a useful tool for combat planners would be a graphical planning tool that could be used simultaneously by a large group of planners. An interactive large screen display could provide the situation awareness and mapping capabilities needed.

In the current manual process, planners sit around a table and have at their disposal a large map of the theater of operations, clear plastic sheets that can be used as map overlays which can be written on with markers, and tabular data containing force status and other relevant information.

An ideal tool would provide information grid overlays. Such a display is illustrated in Figure 8.1. One grid would display the locations of strategic targets. Another grid would display the location of friendly air bases. If an air base icon was "clicked on," a tabular display of attack, tanker, and escort aircraft assets would appear on the screen.

This tool would have a powerful GUI (similar to the potential GUI proposed for FLEX and discussed in Chapter Six). Screen icons would be manipulated to directly estab-

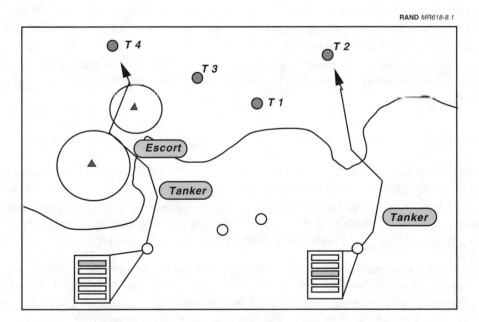

Figure 8.1—Automated MAP Production Tool Display

lish relationships. Icons on the large screen display would be controllable either by mouse or by a pen that planners could use by holding it near the surface of the display. Planners could use the electronic pen to select aircraft at a particular base and could mark the course of the aircraft by dragging the pen along the display. Once the pen reached an appropriate target, the planner could mark the target with the pen, automatically linking an aircraft with a target. A group of aircraft could be selected in this way allowing planners to put packages together quickly and to make package target assignments while standing at the screen. The links between targets and aircraft established by using this tool would be automatically converted into database form for later use in the ATO production process.

The automated MAP production tool would provide indications of when and where in a mission EC escorts or tanker refuelings would be needed (based upon aircraft fuel flows and EOB data). The underlying algorithms used in these computations would intentionally be as simple as possible, like the "rules of thumb" frequently used by MAP planners during manual MAP production. In this way, rough tanker and escort aircraft requirements could be generated in real time as an aircraft course is marked out on the screen. Tankers would be assigned to aircraft in much the same way as attack aircraft would be assigned to targets, by establishing relationships with icons on the screen.

The type of MAP automation tool described above, with a powerful GUI linked directly to underlying databases, would be significantly different in design from RAAP and ACPT. If a modular software design approach is taken, a new tool could be developed that is largely independent of these other systems, but whose databases could be linked automatically to them. However, ACPT or RAAP would be the natural candidates for augmentation with an automated MAP production tool because they are already linked to target databases. In either case, RAAP, ACPT, ICM and the hypothetical MAP automation tool described above would have to be tightly linked to provide the responsiveness needed by planners. The MAP automation tool would have to have immediate access to the master target database so that target and package changes can be made quickly.

RESTRUCTURING CTAPS TARGET AND ORDER OF BATTLE DATABASES

Today there are six separate OB databases in CTAPS 5.0x. They all have to be separately maintained and updated by operators, which could be quite time consuming and difficult during rapidly changing operations. Because of current technology limitations, CTAPS applications that have OB databases are probably best integrated by designating one application as the master OB database controller.

This may be essential because inconsistent OB databases can arise if different applications employ different correlation algorithms in producing their local OB database. If different applications are linked to different information sources, they can also be expected to produce different OB databases. And if different OB update and database linking procedures are used with different applications, one can expect non-synchronized OB databases to result.

These concerns have been recognized, and ICM has been designated as the master OB database for CTAPS. However, additional steps need to be taken to ensure OB database integrity throughout the AOC. A systematic update process should be established to ensure that CTAPS and CIS OB databases are updated in a regular and uniform fashion. Initially, these will probably have to be based on a manual set of procedures.

In the long term, all these databases should be automatically linked and the number of databases reduced to alleviate the work load and time pressure on force-level planners. An integrated database architecture is needed for CTAPS and CIS applications to ensure OB databases are synchronized. To facilitate automatic OB updates, the OB databases of all relevant CTAPS applications should be made accessible to ICM. In this way, ICM could automatically transmit OB updates throughout CTAPS. CTAPS applications that automatically link to ICM could then alert operators of OB changes if they impact current planning activities. Finally, because automatic OB updates could disrupt the planning process, these updates should probably be programmed to occur at preplanned times. That is, automatic OB updates should be scheduled. *If the ATO cycle is to be shortened significantly, a unified CTAPS OB database architecture that electronically links all necessary applications to the ICM master OB database will be essential.*

Target Databases

At least five separate target databases are used in CTAPS 5.0x. Maintaining database integrity and ensuring all target databases are synchronized will also be challenging during combat operations.

In the near term, one database should be designated as the master target database. If changes must be made manually, one duty station should be responsible for transmitting target changes to other database managers, and a clear set of procedures should be established for the target change process. Perhaps RAAP should be assigned the responsibility for maintaining the master target database.[1] Arguments can also be made to suggest that ICM should have this responsibility instead of RAAP. However, ICM will have its "hands full" maintaining OB databases and executing correlation algorithms, so ICM may have less Control Processing Unit (CPU) time available for target database management than RAAP. There are also plans to integrate ACPT with CTAPS. An argument can made for selecting ACPT as the application charged to control the CTAPS master target database. Again we will not belabor this issue here except to say that ACPT would probably have to be operated at the secret level to integrate it to the degree required and to enable it to act as the target database transaction controller.[2] We will not dwell on this issue any further, except to say that it presents an important dilemma. What is the most efficient way to tie together the different target databases used in CTAPS? The solution to this issue will

---

[1] This is in agreement with the RAAP System Specification.

[2] It is also not clear whether downgrading ACPT to the secret level would cause some loss of transparency in the STT "audit trail" which is used to generate target priorities in ACPT.

affect other aspects of the system. For example, which database technologies should be used in future versions of CTAPS? And how can the overall ATO production process be structured? Can certain subprocesses be executed in parallel if they can simultaneously access a master target database?

In the long term the CTAPS database architecture should be modified so all CTAPS target databases are automatically linked to a single master target database, and if possible the number of target databases should be reduced.

Common and Mirror Database Designs

In earlier chapters we described the limitations of the CTAPS 5.0x database architecture and the limited forms of interoperability that exist between the databases used by different CTAPS 5.0x mission applications. These limitations restrict how the current air campaign planning process can be structured and changed. A common CTAPS database design, in which mission applications would share one or more databases that would be automatically updated by CTAPS master databases, would allow CTAPS to be more readily adapted to different air campaign planning processes and would increase the flexibility with which planners and operators could use CTAPS. For example OB databases could be mirrored or replicated in CTAPS. These mirror databases would store data in a common format accessible by all mission applications. A similar set of master and mirror target databases could also be established. This type of database architecture is illustrated in Figure 8.2.

Target, OB, and perhaps published ATO databases will have to be mirrored or replicated in CTAPS. These mirror databases would store data in a common format accessible to all mission applications. Master target and master OB databases would automatically update their mirrors. Other mission applications would access mirror databases to update their own files in order to reduce communications loads on the mission applications maintaining master databases.

Another limitation of the current CTAPS architecture is that a single server provides intra- and inter-application messaging services for all workstations on the network.

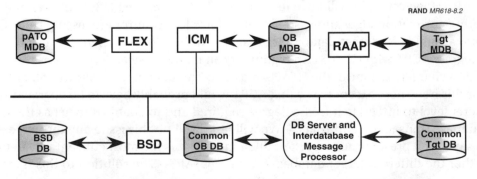

RAND *MR618-8.2*

NOTES: DB: Database, MDB: Master Database, OB: Order of Battle, Tgt: Target, pATO: Published ATO.

Figure 8.2—Notional CTAPS Database Architecture

When large ATOs are being processed in APS, this server becomes a serious bottle-neck to message processing between user terminals and local databases. This short-coming will undoubtedly be fixed in version 6.0 of CTAPS. We just wish to point out here that direct database-to-database transactions, between masters and their mirrors, could be efficiently handled by a dedicated database server and message processor.[3]

The CTAPS 6.0 database architecture will be significantly more complex than the current one and will have to provide real-time access to the published ATO. FLEX will only be as useful as its databases allow. If its database remains current with real events, then FLEX will be able to accurately portray the ATO. A similar statement also holds true concerning the accuracy and utility of the BSD. Both these applications will have to perform a large amount of real-time message processing, and their databases may have to be isolated from other mission applications to provide the level of responsiveness needed.

Another important feature of the CTAPS AOC database architecture should be scalability. Since the number of APS or FLEX workstations will increase, the CTAPS database architecture should be reconfigurable and modular so that additional mirror databases can be added to maintain a constant level of system responsiveness. Thus, as the size of the AOC and ATO increases, the number of replicated or mirror databases can be increased.

Earlier, we proposed that master databases be capable of automatically updating their mirrors according to a preprogrammed schedule. It would be useful for operators to be able to adjust the schedule of database transfers and updates so that they can be adapted to operational circumstances during an air campaign.

PARALLEL TARGET DEVELOPMENT AND WEAPONEERING AND SCHEDULE TARGET CHANGES

The key to reducing ATO cycle time is to divide the subprocesses performed during the cycle into those that can be done in parallel or simultaneously and those that can only be done in series. As discussed in Chapter Three, target development and weaponeering (TD&W) could be performed in parallel with other subprocesses. Some iteration and information sharing would be needed between subprocesses, especially before the target list for the next day's ATO is finalized as the TNL, but otherwise TD&W subprocesses are "parallelizable."

How would TD&W be done independently of other activities? Candidate targets would be developed, weaponeered, and forwarded to combat planners and ACPT. If certain new candidate targets were given a high enough priority, they would be grouped together in an ATO changes file. At preplanned times in the ATO planning cycle all planning activity would stop, placing the ATO database into a static condition. Then the TNL in the APS ATO database could be changed and new high-

---

[3]The detailed design of a more capable message processing system and database architecture for CTAPS is beyond the scope of this investigation.

priority targets added. A similar number of existing targets would be deleted from the TNL consistent with the apportionment guidance from higher authority. By developing and weaponeering candidate targets continuously during the cycle, targeteers could build up a library of fully weaponeered prioritized targets that could be inserted at a suitable juncture into the ATO production process.

An important element of this change in the ATO production process is to introduce *a set schedule for making changes to the ATO*, once actual ATO coordination and deconfliction is under way. This will help coordinate parallel planning activities and will probably be necessary when a large number of changes are made to the ATO. ATO changes would be transferred between CTAPS databases on a prescheduled basis to reduce the number of potential disruptions experienced by ATO planners. However, to carry out these planning activities in parallel, CTAPS target databases have to be electronically linked so the transfer of target changes is made quickly and additional time delays are not introduced into the shortened ATO cycle.

If TD&W is to be done responsively, especially in the case of high-value emergent targets, all CTAPS mission applications involved in the targeting process must have automatic access to a *master or mirror target database*. A portion of this database would be populated by high-priority targets that would be constantly updated and periodically transferred to the APS TNL. All other missions applications needing access to these target databases would be electronically linked to them and automatically receive target updates.

There are technological challenges involved with solving the target database access problem. If all CTAPS applications used the same target database, the number of database transactions may be too large to maintain the level of responsiveness needed. In addition, there are inherent limitations in the designs of current SQL relational databases which make them difficult to link electronically. Object oriented software, including object oriented databases, may offer a way to solve data linkage and inheritance problems. However, this software technology is new and object oriented software development environments and standards have only recently been introduced. Furthermore, this technology may be difficult to adapt it to existing relational database management systems (RDBMSs). It may be possible to automatically link RDBMSs at specified times if all database queries were stopped for a short period of time. A single CTAPS application, for example RAAP, would act as the master target database and would periodically provide a set of changes to other applications or mirror databases. *Such a solution would also be compatible with the modified air campaign planning process proposed in Chapter Three.*

Figure 8.3 illustrates how the ATO cycle could be cut in half if ACPT was used to automate MAP production, if target development and weaponeering were carried out in parallel with other ATO production subprocesses, and if CTAPS databases were linked appropriately. During a 24 hour planning cycle, planners and weaponeers would continuously identify, prioritize, develop, and weaponeer targets. When new high priority targets were identified and weaponeered, they would be added to the ATO changes file, which would be inserted into the MAP or ATO production processes at prescheduled times in the planning cycle.

RAND *MR618-8.3*

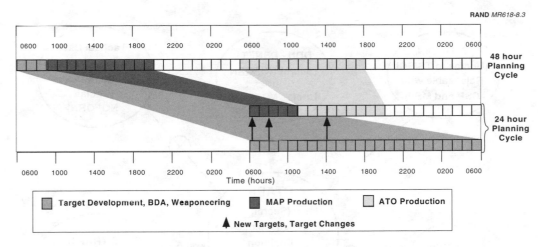

Figure 8.3 —Notional Shortened ATO Cycle

MULTILEVEL SECURITY

Multilevel security in the CTAPS architecture is a larger issue than just how to best integrate ACPT into the system. The general nature of this problem is illustrated in Figure 8.4. From the figure, it is apparent that CTAPS should be able to communicate on networks which operate under a variety of different security rules. Currently, automated MLS interfaces do not exist for connection to any of these networks except for the AUTODIN system. If ACPT is integrated into CTAPS, the MLS connectivity picture will become even more complicated.

As indicated in the figure, the Air Force plans to employ CTAPS in conflicts with coalition partners. While basic CTAPS operating software and hardware will be shared with coalition partners even during peacetime (without OB, targeting, and U.S. aircraft performance parameter database information), it is unclear how operations can be conducted with coalition partners during wartime if coalition partner CTAPS terminals are connected to the AOC LAN (which now operates at the secret/NOFORN security level).

The CTAPS MLS security problem is part of a larger issue of how to provide MLS interfaces to new COTS-based C4I systems. Standard MLS interface products are being developed for TAFIM-compliant C4I systems, but it is unclear how easily they can be integrated with existing GOTS-based messaging systems, or with COTS-based RDBMSs. COTS computer technology continues to advance at a higher rate than DoD MLS technology. This problem concerns commercial vendors of COTS-based computer security products who have asserted that there is a lack of national leadership on this problem.[4] Some vendors have also claimed there has been inadequate government investment in the industry.

---

[4]Uhrig and Robella, 1994.

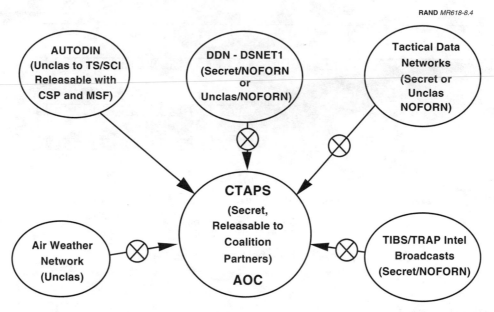

RAND *MR618-8.4*

TS/SCI: Top Secret/Special Compartmented Intelligence.
DDN: Defense Data Network.
DSNET1: Defense Secure Network—1.

SOURCE: Soderholm, 1994.

Figure 8.4—CTAPS Multilevel Security Environment

The lack of DoD MLS systems that conform to commercial computer industry standards has hampered the development of MLS interfaces for client server systems. The recent start of the Multilevel Information Systems Security Initiative (MISSI) will help to remedy this problem. However, more action at the DoD level may be needed. Current policy requires detailed data flow analysis at the data element level to be performed to develop and program MLS interfaces. This activity is difficult and time-consuming to perform, and may have to be performed again when data elements or software application properties are changed. A simpler MLS technical solution is needed so that MLS interfaces can easily be upgraded with system software changes and upgrades. Lack of a more appropriate and flexible MLS policy for COTS-based systems could impede and slow development efforts of systems like CTAPS.

A single set of standard MLS interfaces should be established for TAFIM-compliant client server systems that operate with COTS RDBMSs. MLS standards for client server systems should include and evolve with the joint COE and new messaging standards as these become available.

CTAPS AND ATO INTEROPERABILITY

The ATO will be disseminated to a large number of joint command and control centers and units by means of CTAPS and the GCCS, as shown in Figure 8.5. The figure indicates the large number of information system environments CTAPS and GCCS

RAND *MR618-8.5*

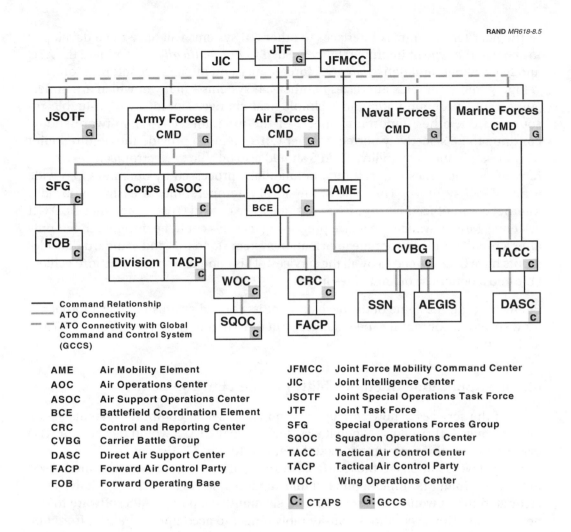

AME — Air Mobility Element
AOC — Air Operations Center
ASOC — Air Support Operations Center
BCE — Battlefield Coordination Element
CRC — Control and Reporting Center
CVBG — Carrier Battle Group
DASC — Direct Air Support Center
FACP — Forward Air Control Party
FOB — Forward Operating Base

JFMCC — Joint Force Mobility Command Center
JIC — Joint Intelligence Center
JSOTF — Joint Special Operations Task Force
JTF — Joint Task Force
SFG — Special Operations Forces Group
SQOC — Squadron Operations Center
TACC — Tactical Air Control Center
TACP — Tactical Air Control Party
WOC — Wing Operations Center

C: CTAPS   **G:** GCCS

Figure 8.5 —Joint ATO Connectivity Supplied by CTAPS and GCCS

will have to be integrated into. The integration of CTAPS in this joint environment is a significant challenge.

The long-term approach to this problem was established when the Assistant Secretary of Defense for Command, Control, Communications, and Intelligence (C3I) mandated that future versions of CTAPS, as well as most other C4I information systems being developed by the services, will transition to the GCCS COE—when the GCCS COE becomes available. The GCCS COE is designed to provide a single common software environment that different service and Joint Command mission applications can run on, even if different types of computers are used. The GCCS COE will also have a number of basic support applications and communications capabilities built in to enable data and messages to be exchanged between different mission applications. The GCCS COE is being developed in parts by the services and will be integrated by the DISA.

In the near term, complex interfaces to other C4I systems will have to be developed to achieve the type of interoperability needed to *automatically* disseminate the ATO and to exchange situation awareness, force status, and intelligence information. Developing many complex interfaces to other C4I systems—many of which are undergoing rapid programmatic and technological change—could divert valuable resources from other important aspects of the program. CTAPS software should continue to be selectively ported to other C4I systems to provide the required joint ATO dissemination capability. CTAPS should also be able to electronically exchange raw information files that can then be manually processed by operators in a wide array of C4I systems. This type of low-level interoperability is easily achieved if internet communications protocols continue to be used in CTAPS and other DoD C4I systems. More advanced message processing features could be developed, but only after a broad set of appropriate information system standards and standard data elements have been agreed to by all the services. This approach should be more achievable and a better use of limited resources.

Below we examine some of the development issues and promising new approaches for distributed computing and database structures that may play a role in the CTAPS architecture.

Mission Application Interoperability in a Joint Environment

Each of the services is now employing open client server computer systems as the basis for new C4I systems. When it was decided that CTAPS would be used for ATO dissemination, the Navy determined that it had to port CTAPS version 5.0 software from SUN workstations to the Hewlett-Packard (HP) workstation environment used in the Navy Joint Maritime Command Information Systems (JMCIS).[5] Initially, it was estimated that it would take from three to six months to port CTAPS software to this new system. However, it took considerably longer to accomplish the port from one platform to another. In the intervening time, numerous bugs were discovered in initial versions of CTAPS 5.0x software. Patches were devised to eliminate these bugs, but the frequent revision of CTAPS version 5.0x software has made it difficult to maintain full system software interoperability with the Navy.

The Air Force now plans to release new versions of CTAPS software every 18 months, and to provide backward compatibility to one previous version of the software build.[6] This upgrade strategy will ensure interoperability as long as no unanticipated incompatibilities emerge between chronological and ported versions of the software. However, in recent exercises, serious interoperability problems were experienced between a ported version of CTAPS software and Air Force CTAPS systems.[7] The problem resulted because a Navy ship had not yet been upgraded with the latest

---

[5]It was necessary to port CTAPS software. If a CTAPS terminal was connected as a client system to an HP client server network, the CTAPS terminal would cause serious disruption of Navy LAN communications.

[6]*CTAPS Update Briefing to the Theater Battle Management 06 Advisory Working Group Meeting,* Col. Carl Steiling, ESC/AVB, July 1994.

[7]These problems were experienced during Exercise Tempo Brave 94-1 between CTAPS systems operated by the 13th Air Force and two CTAPS terminals used on the U.S.S. Blue Ridge.

ported version of CTAPS, even though it had been available for some time. Navy ships usually do not upgrade their systems while at sea, and the new version had become available while this ship was deployed.

Operational differences and different system upgrade strategies may make it difficult to maintain CTAPS interoperability at all times between Air Force and Navy units. And as CTAPS software becomes more complex, it may be more difficult to port software from one vendor platform to another. Therefore, a software development strategy must be devised that reduces the probability of unanticipated interoperability problems emerging from the porting process, and which can reduce the difficulty in porting software from one open system vendor platform to another. The GCCS COE, if properly structured, can help address both of these concerns.

The GCCS COE is depicted in Figure 8.6. As indicated it will provide the basic support and application services in GCCS and other systems based on it to help ensure interoperability. The COE is based on TAFIM standards but is more detailed in that it describes needed standards for higher-level support and application services. It will provide a standard set of application program interfaces that all mission applications can use. The set of services the COE provides will be reconfigurable and tailorable to support a wide range of tactical C4I system needs. The COE will be based as much as possible on COTS products.

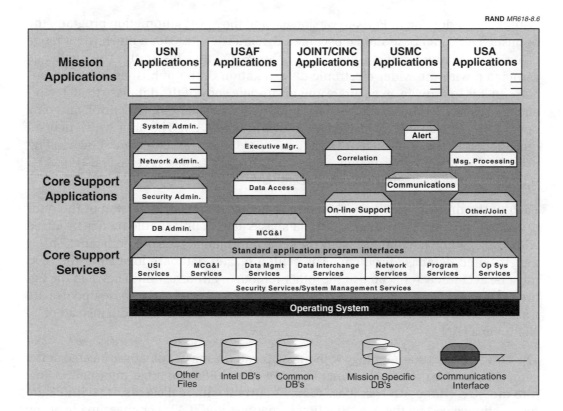

SOURCE: Interservice Common Operating Environment Working Group, 1994.

Figure 8.6—Common Operating Environment

The COE working group is using an evolutionary standards setting process. It will be a "living" set of standards. As new standards emerge for useful services they will be added to the COE. The COE also calls for establishing standards for certain services where none exist at present. As the required standards are developed they will be added as well.

The COE will be developed so it can be compiled and used on a variety of hardware platforms. This will allow each service to contract independently for open system hardware. It is proposed however that contract awards be reviewed by an Interservice Configuration Control Board to ensure they conform to COE standards.

A developer's tool kit for the COE will also be developed that can be used to add new features to COE-based systems. It will permit systems developers to add new application program interfaces and other functionality to the system while adhering to COE information system standards. It will include a data dictionary as well.

With this approach, it should be possible to quickly port, in a reliable way, new versions of CTAPS software to other open systems which will be characterized by their own dynamic COE-based software environments.

Real-Time Joint Access to the ATO Database

With the introduction of FLEX in version 6.0 and the ASOC automation program, the published ATO database should become a dynamic database for a much wider range of units and commanders—not just for the JFACC and his staff. The challenge associated with the wider electronic dissemination of the published ATO and updates to it is to provide an accurate and up-to-date copy of the database to all recipients and especially those who need the information to carry out, or control, on-going operations (for example an ACE on AWACS or Joint STARS). The units which may require access to the published ATO while the ATO is being executed are shown in Figure 8.7. Today the airborne units shown in the figure cannot receive real-time ATO updates, except by means of communications.

Each of the units in the figure should have access to the latest version of the published ATO and to ATO changes. ATO databases at each unit should be synchronized with one another, or contain exactly the same information to avoid problems. This is a distributed database problem. The central master database located at the AOC must be accurately replicated at the other remote or airborne units shown in the figure. Database mirroring is more difficult in this case because database updates are transmitted over WAN communications media with limited bandwidth instead of locally over a LAN.

The network has a star topology with the AOC and the master database located at the center. Database updates flow out of the AOC. MISREPS and other information flow into the AOC. If the AOC database remains current with real events, if MISREPS are filed and received by the AOC in a timely manner, and if ATO changes made at the ASOC are communicated to the AOC quickly, then FLEX will provide the units in the figure with an accurate picture of the ATO.

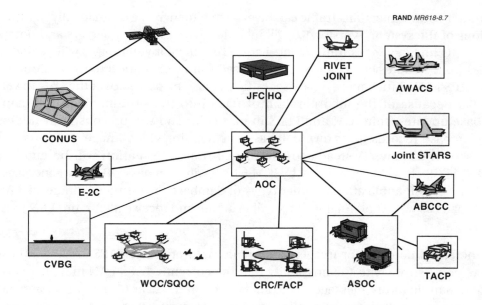

Figure 8.7—Replicated Dynamic ATO Databases

In practice, during a real operation it will be difficult to keep all the databases shown in the figure synchronized. They will frequently be inconsistent to some degree because of communications delays and processing bottlenecks. A robust distributed database and WAN architecture is needed to cope with these real world effects.

Obtaining synchronization between a master database and its mirrors is difficult to do. There are serious technical hurdles in providing this capability in a reliable way using present day RDBMSs. One approach has recently been suggested by RAND for another application by using knowledge-based systems.[8] This approach may be especially well suited for maintaining OB databases but could also be applied for the problem of maintaining currency of the published ATO with real events.

New object oriented database management systems can attach life cycle and historical attributes to objects and can track objects by using inheritance features. These capabilities may also enable innovative strategies to be employed for these database update and synchronization problems.

CTAPS Local Area Networks

CTAPS was originally designed for operation exclusively in a high-capacity LAN environment (the CTAPS LAN was originally intended to be a fiber-distributed data interface system which would provide 100 Mbps of bandwidth). Currently Ethernet is used, and in large client server systems like CTAPS, an Ethernet LAN can become heavily loaded with communications traffic, especially if large files and databases are transferred regularly.

---

[8]Kameny, September 1994.

Computer communications traffic can be expected to increase dramatically in future versions of the system. Version 6.0 will include CIO 5D imagery servers and potentially a CIO imagery mensuration tool. Large (100s of kilobyte) images will be downloaded from remote servers and may be passed from workstation to workstation on the CTAPS LAN. Much more database traffic will also be passed over the LAN in version 6.0 because of the incorporation of JDSS into the system. JDSS will copy database updates from ICM and FLEX in particular and will be constantly adding new database records to its own database and sending video image elements to a large screen display. A steady stream of real-time information will be entering CTAPS via CS. It will be processed by ICM which will then pass OB update messages to other mission applications. Other types of database mirroring will increase LAN traffic loads as well. And finally, FLEX will generate and receive a lot more LAN message traffic.

The increased demand for imagery alone may saturate the CTAPS LAN. All these factors point strongly to the possibility that the current CTAPS LAN may be inadequate to handle future message traffic loads, even in the near term with version 6.0. In the far term, if multimedia applications and messaging were to be incorporated into CTAPS, network loading would increase even more dramatically.

The Air Force needs to develop a long-term acquisition plan that includes standards for high-capacity LANs at both the force and unit level. LAN-based asynchronous transfer mode switching systems may not be mature enough, and asynchronous transfer mode standards for LANs probably will not be available by 1996 to ensure interoperability among vendor systems. The Air Force must give serious thought to establishing standards for advanced Ethernet-based LANs in the near term.

In the long term, the Air Force and DoD need to reassess R&D priorities for advanced networking systems. Such R&D efforts should be refocused on tactical warfighter needs, and on developing moderate bandwidth standardized LANs with capacities between 100 and 500 Mbps.

CTAPS Wide Area Networks

As mentioned above, CTAPS was originally designed for a high-capacity LAN environment. In the current version of the system, the CTAPS LAN can be heavily loaded with a variety of computer communications traffic. CTAPS was designed for a "communications rich, computer poor" environment. In version 5.0x, CTAPS servers query other workstations on the network , using so-called daemon processes, to find out if they are busy. If some workstations have free processor time, the server will transfer computer jobs to other workstations so as to balance the computational load evenly throughout the entire client server system. In addition, a relatively large number of SQL messages are sent between client and server machines when a single change is made to a relational database. The COTS-based RDBMSs used in CTAPS were also originally designed for a communications-rich environment.

These CTAPS daemon and SQL processes consume a significant amount of communications bandwidth. In a tactical wide area communications environment they can lead to serious performance degradations, for example when querying an AOC database from a remote CTAPS terminal. In recent exercises observed by RAND, relatively simple remote database queries sometimes took a long time to execute. In these exercises the ATO was rarely transferred in a timely fashion by using CAFMS. The ATO transfer problem could have been due to a number of different causes, including operator error.[9] However, overall CTAPS WAN performance would still be improved by reducing the number of extraneous background processes on the WAN.

The CTAPS daemon process implementation in version 5.0x, originally designed for sharing local computing tasks, should be easy to modify to reduce background processes on the CTAPS WAN. This may have already been done but should certainly be addressed before version 6.0 is released.

New COTS technologies can also improve CTAPS WAN performance and reduce SQL message traffic generated during database queries. Oracle, a major vendor of RDBMSs and the maker of one of the RDBMSs used in CTAPS, has developed a new database interface designed especially for use over wireless or WAN links. The average database query on a client server system takes about ten sets of "round trip" message pairs, with each round trip taking about 20 seconds.[10] The new wireless database product reduces the number of intermediate messages sent over the WAN and increases the overall responsiveness of remote database queries.

The above discussion indicates that complex technical issues are involved in ensuring that the CTAPS WAN will perform properly in a conflict. Because CTAPS software and hardware, and other C4I systems that CTAPS will interoperate with, will change frequently, it is nearly impossible to predict how well or poorly the system will perform. In addition, CTAPS could be used in a wide variety of operations, in many different types of WAN configurations, and CTAPS WAN underlying communications paths could be supplied by a number of different systems. These factors will also make it difficult to predict CTAPS WAN performance.

CTAPS should therefore be tested in a wide range of configurations, using a variety of scenarios including some with large ATOs, and some with a large number remote CTAPS terminals. Detailed CTAPS WAN performance monitoring should be done during exercises and stress tests to determine the cause of performance degradations that may occur.

An institutional mechanism may be needed in the Air Force R&D community to experiment with new COTS networking and communications technologies, and to gradually incorporate new prototype systems into existing tactical networks. The Electronic Systems Command Technical Interoperability Network is a start in this direction, and could be broadened to include connectivity with Rome Laboratory,

---

[9]The ATO transfer problem occurred between the AOC operated by the Thirteenth Air Force on Guam and the USS Blue Ridge during Exercise Tempo Brave 94-1. Computer system incompatibilities on board the Blue Ridge, system setup errors on the ship or at the AOC, other system usage errors, or satellite communications problems could have caused some of the ATO interoperability problems experienced during the exercise.

[10]If satellite links are used, these time delays are probably larger.

other joint labs, and with selected operator locations. The Technical Interoperability Network would be one component of a new evolutionary acquisition in which new systems could be tested in a realistic network environment.

Finally, because of the increasing jointness of the program, CTAPS must be considered a distributed system designed to operate in a joint WAN environment. Data communications requirements for CTAPS need to be folded into a single database of joint WAN requirements since these networks are frequently shared with other users.

Borenstein, Nathaniel, *MIME: A Portable and Robust Multimedia Format for Internet Mail*, Multimedia Systems, p. 29, Springer-Verlag, 1993.

Bouchoux, D., *Briefing: Air Tasking Order (ATO) Generation and Transmission*, MITRE, 20 July 1992.

Briefing: Contingency Theater Air Control System Automated Planning System (CTAPS), 24 February 1993.

Clark, Thomas A., Georganne H. deWalder, Earl C. LaBatt, and Gerald C. Ruigrok, *Force Level Execution (FLEX): Developing and Demonstrating a Monitoring and Execution Capability for Combat Operations*, Rome Laboratory, Griffiss AFB, NY, June 1994.

Cohen, Eliot, et al., *Gulf War Air Power Survey, Vol. I: Planning and Command and Control*, Office of the Secretary of the Air Force, 1993, SECRET/NOFORN/WNINTEL/NOCONTRACT.

Commander in Chief, U.S. Pacific Forces, "Exercise Tempo Brave 94-1 Planning Documents," n.d.

CTAPS Bidders Library, *Overview of the Battlefield Situation Display for CTAPS*, Reference No. 171, n.d.

CTAPS: The Leading Edge of the U.S. Air Force Theater Battle Management Philosophy, n.d.

Defense Information Systems Agency, Center for Standards, *Department of Defense Technical Architecture Framework for Information Management*, Vol. 7, Version 2.0, Adopted Information Technology Standards (AITS), 12 May 1994.

Defense Information Systems Agency, Center for Information System Security, *Department of Defense Technical Architecture Framework for Information Management*, Final Draft, Vol. 6, Version 1.0, 1 August 1993.

Department of the Air Force, PMD2328(1)/27438F/41840F/63617F/62322D/27412F, *Program Management Directive for Theater Battle Management (TBM) Core Command and Control (C2) Systems*, Office of the Assistant Secretary, Washington D.C., 18 September 1993.

Endicott, Don, *The Defense Research and Engineering Network (DREN)*, DREN Program Overview, 19 May 1994.

Griffiss AFB, *Advanced Planning System (APS) Operator Training Manual*, Rome Laboratory, NY, n.d.(a).

Griffiss AFB, *Point Paper on Force Level Execution (FLEX) AWACS Version*, Rome Laboratory, NY, 25 July 1994a.

Griffiss AFB, *Statement of Work for Force Level Execution*, Rome Laboratory, NY, PR No. C-4-2104, 14 January 1994b.

Griffiss AFB, *Theater Battle Management: Integrated C3I Technologies*, Rome Laboratory, NY, n.d.(b).

Interservice Common Operating Environment Working Group, *White Paper on the Common Operating Environment*, draft, 19 May 1994.

Joe, Leland T., and Daniel Gonzales, *Command, Control, Communications, and Intelligence Support of Air Operations in Desert Storm* (U), N-3610/4-AF, RAND, 1994, SECRET/ NOFORN/WNINTEL.

Kameny, Iris, et al., *DIS Need for DoD Data Standards*, Memorandum for Danette Haworth, 11th Workshop on the Interoperability of Defense Simulations, Paper 94-123, 29 August 1994.

Kameny, Iris, *Information Standards Issues Relevant to Army C3I Architecture*, 8 March 1993.

Kameny, Iris, *An Approach to Replicated Data Bases for Robust C2*, DRR-847-A, RAND, Santa Monica, CA, September 1994.

Kirks, David, LTC, *Army Global Command and Control System*, PM SACCS, Ft Belvoir, VA, 17 May 1994.

LaBatt, Earl C., Jr., *Marquee: A Graphical Mission Monitoring Tool*, RL/C3AA, Griffiss AFB, NY, n.d.

Object Management Group and X Open, *The Common Object Request Broker: Architecture and Specification*, John Wiley & Son, Inc., NY, OMG Document Number 91.12.1, 1992.

Object Management Group, *Object Technology: A Component for the Future*, white paper, Framingham, MA, n.d.

"Oracle to Roll Out Wireless Links," *PC Week*, 12 September 1994.

Paige, Emmet, Assistant Secretary of Defense for C3I Memorandum, *Development or Modernization of Automated Information Systems (AISs)*, 8 December 1993.

Paige, Emmet, Assistant Secretary of Defense for C3I Memorandum, *Selection of Migration Systems*, 8 December 1993.

Papagni, J., *Point Paper on Rapid Application of Air Power (RAAP)*, Rome Laboratory, Griffiss AFB, NY, 28 June 1993.

Perry, Walter, Deputy Secretary of Defense Memorandum, *Accelerated Implementation of Migration Systems, Data Standards, and Process Improvement*, 13 October 1993.

Powell, Colin, General, the Joint Chiefs of Staff, *C4I for the Warrior*, June 1992.

Prowse, Mike, LTC, *Briefing: Sentinel Byte*, Intelligence Data Handling System Program Office, USAF, n.d.

Ruigrok, Gerald C., *Force Level Execution (FLEX): A Case Study for a Flexible User Interface*, Rome Laboratory, C3AA, Griffiss AFB, NY, n.d.

SAIC, *Contingency TACS Automated Planning System (CTAPS)*, briefing, 22 March 1993.

SAIC, *Contingency Theater Air Control System Automated Planning System (CTAPS)*, CTAPS V5.0, System Administrator Training for 5th Air Force, 93-09-001, 10 September 1993.

Scott, Karyl, *Open Systems Today*, "OpenDoc: New Take on Object Technology," 1 August 1994.

Soderholm, Steve, Captain, *CTAPS Communicatins Connectivity of Coalition Warfare*, Air Combat Command/Directorate of Combat Requirements, briefing, 26 July 1994.

Soley, Richard Mark, *OMG: Creating Consensus in Object Technology*, Object Management Group, Inc., Framingham, MA, 1993.

Steiling, Carl, Colonel, *Overview of Contingency Theater Automated Planning System (CTAPS)*, Electronic Systems Center, Hanscom AFB, MA, n.d.

Tanenbaum, Andrew S., *Computer Networks*, Second Edition, Englewood Cliffs, New Jersey: Prentice Hall, 1989.

Thaler, David E., and David A. Shlapak, *Perspectives on Theater Air Campaign Planning*, MR-515-AF, RAND, Santa Monica, CA, 1995.

Uhrig, Rick, and Barry Robella, *Multilevel Security Database Management Systems*, Sybase, 8 June 1994.

USAF, Electronic Systems Division, *Briefing: Constant Source*, 17 September 1990.

USAF Scientific Advisory Board, Report of the Ad Hoc Committee, *Information Architectures That Enhance Operational Capability in Peacetime and Wartime*, USAF, SAB 94-002, February 1994a.

USAF, Tactical Fighter Weapons Center, Tactical Analysis Bulletin, Vol. 91-2, July 1991, SECRET/NOFORN/WNINTEL/NOCONTRACT.

USAF, Commander in Chief, U.S. Pacific Command, U.S. Atlantic Command, *Joint Force Air Component Commander (JFACC) Concepts of Operations*, J311 3000 Ser 7, 15 January 1993.

USAF, Deputy Chief of Staff, Plans and Operations Headquarters, *JFACC Primer*, Second Edition, February 1994b.

USAF, *The Air Campaign Planning Tool (ACPT)*, briefing, Air Force Deputy Chief of Staff Plans and Operations, Directorate of Operations, CHECKMATE Division, n.d.

Whitehurst, Charles M., LTC, *Joint Air Tasking Order Interoperability: A Contemporary Review*, Maxwell AFB, AL, June 1993.